Entanglements

Entanglements

Conversations on the Human Traces of Science, Technology, and Sound

Simone Tosoni with Trevor Pinch

The MIT Press
Cambridge, Massachusetts
London, England

This book was set in Stone Serif and Stone Sans by Toppan Best-set Premedia Limited. Printed and bound in the United States of America.

Library of Congress Cataloging-in-Publication Data

Names: Tosoni, Simone. Works. Selections. | Pinch, Trevor, 1952- Works. Selections.
Title: Entanglements : conversations on the human traces of science, technology, and sound / Simone Tosoni and Trevor Pinch.
Description: Cambridge, MA : The MIT Press, [2016] | Includes bibliographical references and index.
Identifiers: LCCN 2016018211 | ISBN 9780262035279 (hardcover : alk. paper)
Subjects: LCSH: Technology--Social aspects. | Human engineering.
Classification: LCC T14.5 .E57 2016 | DDC 303.48/3--dc23
LC record available at https://lccn.loc.gov/2016018211

10 9 8 7 6 5 4 3 2 1

Contents

Acknowledgments

I would like to thank Matteo Tarantino, Massimiliano Guareschi, Christine Leuenberger, Paola Sorrentino, Dario Marsic, Cesare Silla, and Davide Lampugnani for their invaluable contribution to the realization of this book, and Harry Collins for his inestimable remarks on the manuscript. A special thank you to Trevor Pinch for his commitment and dedication to this project.

Introduction: A Five-Step Guided Tour of the Social Construction of Technology

Most of what we do today is, in full or in part, mediated by technology. Therefore, science and technology studies (STS) have become a precious—and hardly negligible—interlocutor for any research field studying contemporary social life, both in its mundane and less ordinary aspects. With varying promptness, scholars within fields as diverse as urban studies, mobility studies, organizational studies, body culture studies, media studies, visual and sound studies, and others have been engaging in an increasingly systematic dialogue with STS. This allows them to enrich their own take on their specialized objects of research and to better account for their manifold relationships with technological mediation. At the same time, scholars with a background in STS are participating directly in these fields, often with leading roles.

Yet, to a newcomer, the first exposure to STS may be bewildering. Notwithstanding the growing number of handbooks, manuals, and introductions published in the present phase of academic institutionalization of the field, it may be not easy to find a first orientation within a literature that during the last thirty years has become overwhelmingly large. In particular, a newcomer could be disoriented by the variety of approaches, conceptual frameworks, methodological stances, and philosophical assumptions that represent the interdisciplinary richness of STS and of its often-inflamed internal debates.

Getting an adequate introduction to the social construction of technology (SCOT) can be even more complicated. Introduced in 1984 by Trevor Pinch and Wieber Bijker[1] in a groundbreaking article that launched the

---

1. Pinch and Bijker 1984, reprinted in Bijker, Hughes, and Pinch 1987.

whole program of the new sociology of technology, SCOT represents one of the most prominent approaches within STS and one of the most influential for the neighboring research fields. As will be illustrated in detail in *Entanglements*,[2] in that early paper the authors revealed how technological artifacts are prone to the *interpretative flexibility* of different *relevant social groups*, and how this flexibility reaches a state of *stabilization* to shape the artifact itself. In this way, SCOT demonstrated the relevance of social factors in modeling technological objects, opening "the black box" (to quote an early passage from R. Whitley[3] that became somewhat of a motto) of technological innovation and exposing the shortcomings of technological determinism.

Yet, rather than proposing an all-encompassing and ready-made model to describe the relationship between technology and society, *The Social Construction of Facts and Artifacts* aimed to define the main guidelines of a new research program to come. Over thirty years of empirical research within this program have contributed to SCOT's development, refinement, and even revision. This scholarship has been published mainly as book chapters or papers in leading STS and sociology journals (*Social Studies of Science, Technology & Culture; Science, Technology & Human Values; Qualitative Sociology; Sociology; The Sociological Review*; and *Theory and Society*). Additionally, these papers have appeared in journals belonging to such diverse fields as organizational studies (*Organization Studies*,[4] *Management*),[5] discourse analysis (*Discourse & Society*),[6] media studies (*First Monday*),[7] economics (*Cambridge Journal of Economics*),[8] sound studies (*Journal of Sonic Studies*),[9] and medical studies (*British Medical Journal*).[10] To make matters even more complicated, these applications and eventual updates of SCOT have never been systematized into a monograph. It has in fact been twenty years since

2. See chapter 3.
3. See Whitley 1972: "An ideology of 'black boxism' … restricts research to the study of currently observable inputs to, and outputs from, a system. Any study of the internal processes, which may be unobservable at the moment, is declared taboo" (63).
4. Darr and Pinch 2013.
5. Jarzabkowski and Pinch 2013.
6. Clark, Drew, and Pinch 1994, 2003.
7. David and Pinch 2006.
8. Bijker 2010; Pinch 2010a.
9. Pinch 2014b.
10. Bal, Bijker, and Hendriks 2004.

Wiebe Bijker summarized the approach using his original case studies of bicycles, bulbs, and Bakelites.[11]

In this broad and dispersive scenario, anyone wishing to get an exhaustive overview of SCOT will encounter quite some challenges, the biggest of which is having to chase scholarship left and right (often in unexpected places) without much of a map. This book intends to provide some help to this reader.

Another challenge for this lost reader is genealogical. As becomes apparent very early in studying the paradigm, SCOT cannot be fully understood without at least a smattering of the Empirical Programme of Relativism (EPOR), developed from the mid-1970s by the Bath School within the field of the sociology of scientific knowledge (SSK). With *The Social Construction of Facts and Artifacts*, Pinch and Bijker adapted EPOR, which had originally been conceived to study the social construction of scientific *facts*, to study also the social construction of technological *artifacts*. Throughout its evolution, SCOT has always intended to stay true to this notion of integrating the study of science and technology. The approach can be fully understood only in this light. Therefore, the next chapters will attempt to familiarize the reader with the main insights and debates of the Bath School.

Our lost and challenged reader is, anyhow, not alone. Her difficulties in gaining a comprehensive understanding of SCOT are by and large visible in the ongoing cross-disciplinary debates. Much of the literature is apparently stuck somewhere between the 1980s and the early 1990s: outside the specialized field of STS, the early formulations of the approach appear to be known, discussed, and criticized more than its recent developments.

I write from the perspective of media studies, my own academic field, which represents an egregious example of this temporal displacement. A systematic dialogue between media studies and STS can be traced back to the early 1990s. At that time, British and European cultural audience studies found in the social constructivist school, and especially in SCOT, the theoretical and methodological ammunition they needed to confront a technological determinism that, in some forms, was still influential in media studies.[12] Cultural audience studies deal with how the ways we

11. Bijker 1995. See also Bijker 2015.
12. Silverstone and Hirsch 1992; Mackay and Gillespie 1992; Silverstone 1994; Jackson, Poole, and Kuhn 2002, Bakardjieva 2005. For an overview of the ongoing debate between the STS and media studies, see Lievrouw 2002, 2014; Boczkowski and Lievrouw 2007; Couldry 2008; Wajcman and Jones 2012.

engage and use media contribute to shape our everyday life and, at the same time, with how our daily practices and routines shape different forms of media usage. Hence, media are prone to *interpretative flexibility* like any other technological artifacts. From this standpoint, the social representations of different *social groups* within media audiences and users are central; they are, in fact, deeply connected to specific and *stabilized* forms of media engagement that ultimately mediate the social effects of media. As stated by Boczkowski and Leah Lievrouw, "The rejection of technological determinism, and the acceptance of a relatively strong form of social constructionism, has become the prevailing perspective in new media studies in Europe, North America, and elsewhere. This development can be counted as one of the most important cross-disciplinary influences of STS on the field."[13] Notwithstanding its pivotal role in this revolution, SCOT has been, and still is, mainly addressed in the 1984 formulation.

This book represents an extended dialogue with Trevor Pinch, and it was built with the intention of helping both newcomers and scholars interested in technological mediation with a gateway to the SCOT approach and, through it, to the broader STS field.[14] *Entanglements* was derived from the integral transcription of four rounds of conversation between Trevor Pinch and me, each lasting about two hours, recorded between September 2012 and October 2014 in Ithaca (NY), Paris, and Milan. The four sessions have been revised for coherence and readability, integrated with supplementary material from epistolary exchanges, and edited into four chronological chapters. The chapters follow Pinch's work and career from the early years in SSK to his most recent works on selling and sound. The discussion about SCOT is therefore interposed with recollections of the climate of the STS field throughout the years, from the enthusiasm of its first steps to the present phase of academic institutionalization.

In addressing the most recent development of SCOT, the book focuses also on the theoretical and methodological issues central to the cross-disciplinary dialogue between STS and neighboring fields—especially, and inevitably, with media studies. In particular, throughout the text, *Entanglements* will address five main issues in depth: (1) relevant social groups, (2)

13. Boczkowski and Lievrouw 2007, 954.
14. For another academic interview, focusing on the "convergences and differences between … [STS] and both cultural sociology and cultural studies" (64), see Magaudda 2014.

the intertwinement of social representations and practices, (3) tacit knowledge, (4) nonhuman agency, and (5) SCOT's take on politics.

Concerning *relevant social groups* (1), the interview will describe the elaborate map of social groups addressed by SCOT throughout its development. The focus on groups of users (and nonusers) of technological artifacts typical of the 1984 application of SCOT has been progressively enriched to include not only developers and producers, but also understudied social actors, such as testers or sellers. Moreover, the conversation will address the ways in which SCOT has framed the relationship between technological artifacts and social actors. On the one hand, throughout SCOT's evolution, this relationship has been increasingly addressed as one of mutual shaping: social actors are transformed by the involvement of technological artifacts in their practices as much as they transform technology. On the other, the approach has focused more on the active role of users in shaping the technological artifact, describing, for example, how specific social representations are connected to practices of customization.

In SCOT, as in the Empirical Programme of Relativism in SSK from which it stems, social representations are in fact inextricably connected to social practices: how a technological artifact is experienced, conceived, and described is deeply related to what is actually *done* (or would be done, in case of nonusers) with that artifact. The marked emphasis on representations typical of the first formulation of SCOT has somehow overshadowed the attention the approach has actually dedicated to this interconnection, inducing some scholars to oppose SCOT to more practice-centered approaches. To clarify this misunderstanding, the *intertwinement of social representations and practices* (2) will represent a second focus of the interview.

The relationship between performativity and materiality is of paramount importance to fully understand social practices. In *Entanglements*, this node will be addressed from two perspectives. The first is a focus on its nonrepresentational aspects (e.g., affects, embodiment, and bodily memory), which have been recently reemphasized by nonrepresentational theories within human geography. Once again, the early SCOT's accent on social representations has overshadowed how the social constructivist tradition has drawn on the concept of *tacit knowledge* to develop its fully fledged approach to the *nonrepresentational* (3).

The second perspective is a discussion of the topic of materiality from the perspective of *nonhuman agency* (4). As it will be retraced in *Entanglements*, the issue of nonhuman agency has prompted one of the longer-lasting contrapositions within the fields of SSK and STS: scholars inspired by Actor-Network Theory (ANT) attribute a capability to act to nonhumans that scholars inspired by the Bath School and by SCOT refuse to concede. Although in recent years this contraposition has softened (with Bruno Latour proposing a narrower definition of agency[15] and Trevor Pinch conceding a restricted form of agency to nonhumans),[16] the two stances remain distinct and divergent. Outside the field of STS, the first position seems to prevail within those approaches that stress the theoretical relevance of materiality. In current media studies, for example, recent attempts to account for the role of materiality in embodied media engagement tend to assume, in a taken-for-granted way, nonhuman agency as a theoretical and methodological prerequisite. On the contrary, the centrality of materiality for SCOT does not imply the attribution of agency to nonhumans. The interview will clarify how, notwithstanding some polemical radicalization, the reasons behind this position are, above all else, methodological. The attribution (or the refusal) of agency to artifacts is, in fact, regarded by SCOT as an *ontological* claim, which falls outside the duties (and the possibilities) of a sociologist. From this perspective, in fact, ontology (what something is) cannot be distinguished from epistemology (how something is known): to avoid this short circuit, the sociologist should rather take a step back and account for *who* is attributing agency to nonhumans in a given situation, for the *kind of agency* that is being attributed, for the *contextual conditions* of this attribution, and for its *consequences*. The recently introduced notion of *material scripts* will be discussed, together with the related *accomplishing approach to sociomateriality*, as a methodological tool to achieve this goal: a goal that, from the perspective of SCOT, would be contrarily hindered by a preliminary attribution of agency to material artifacts.

The form of methodological relativism just described has drawn on SCOT several accusations of political blindness or disengagement. As a fifth

---

15. See Latour and Venn 2002; Sayes 2014.
16. See Pinch 2015a: "Sound, like wine can be a source of surprise. This means some sort of agency must be given the material world in how a sound is experienced. No matter how hard or brilliantly a saxophone is blown, it is not easy to make the sound of the Moog electronic music synthesizer. Materiality matters even if it is simply in resisting our ambitions for it (Pickering 1985). A Moog synthesizer, however well played, cannot easily make the sound of the Niagara Falls in full flight anymore than cheap plonk can easily be made to taste like a top bottle of Latour Burgundy" (17).

and final topic, the interview will focus on the specific *take on politics* (5) implied in SCOT.

In order to serve as an extensive introduction, this five-step guided tour within the SCOT approach has been annotated with bibliographic references, clarifications, and suggestions for further reading. Although the final version of the conversations has been reviewed, amended, and approved by Trevor Pinch, I am solely responsible for any omissions or inaccuracies. The interviews were conducted in informal settings and, although Professor Pinch was aware of the general arc of the inquiry, no specific questions were screened in advance. Thus, his answers are best read as part of a conversation. They are what conversation analysts would call "occasioned responses." The textual references in the footnotes provide a complementary route into the debates, detailing the many nuanced positions and carefully considered shifts in the arguments to be found in the literature.

ST

1 The Sociology of Scientific Knowledge

1.1 The Early Years and the Edinburgh Strong Programme

Simone Tosoni I think that the best way to go through the intellectual journey of the social construction of technology [SCOT] approach is to start from its roots in the early sociology of scientific knowledge. To properly understand what you and others were doing in those years, I would start from the 1970s, when several scholars in the history and philosophy of science had already started to deconstruct a canonical and positivistic view of science. I am thinking about Feyerabend,[1] with his attack on the assumed

1. See Feyerabend 1993 (1st ed., 1975). As Feyerabend himself summarized in the introduction of the Chinese edition of *Against Method*, "the thesis [of the book] is: The events, procedures and results that constitute the sciences have no common structure" (1993, 1), science being an "essentially anarchic enterprise" (9). Feyerabend underlined the mismatch between the actual practices of "doing science" and their philosophical and methodological representations; he rejected the assumed distinction between norms and facts, observational terms and theoretical terms (being a "fact" always-already coconstructed by the "theory" from within which it is observed) as well as between a context of discovery (that "may be irrational and need not follow any recognized methods" (147) and that should be addressed by the history of science) and a context of justification (domain of a standardized and prescriptive methodology, addressed by the philosophy of science). As a consequence, Feyerabend held that "far from being irrelevant for the standards of test" (148), the attempt "to retrace the historical origins, the psychological genesis and development, the socio-political-economic conditions for the acceptance or rejection of scientific theories" (148), advocating for an anthropology of science and opening the field to a sociology of scientific knowledge to come: "Such an inquiry ... will have to explore the way in which scientists actually deal with their surroundings, it will have to examine the actual shape of their product, viz. 'knowledge' and the way in which this product changes as a result of decisions and actions in complex social and material conditions. In a word, such an inquiry will have to be anthropological" (197).

role of a prescriptive method in science, or Kuhn,[2] with his dismissal of a linear and cumulative idea of science. Furthermore, at that time there already was a full-fledged sociological approach to science, the functionalist approach by Robert Merton,[3] but it did not seem to be fully adequate in addressing the new idea of science that was taking shape from these attempts at deconstruction. Would you describe this early intellectual context and its relationships with the new sociology of science?[4]

Trevor Pinch   Yeah, okay. So, I think that one contrast between then and now exactly as you said: the philosophy of science was interested in general epistemological issues across the sciences. So, Karl Popper's

2. See Kuhn 1970 (1st ed., 1962). Kuhn's historical work played a key role in dismissing the idea of scientific progress as linear and cumulative. Drawing on his renowned concept of paradigm ("universally recognized scientific achievements that for a time provide model problems and solutions to a community of practitioners," 10), Kuhn proposed instead a depiction of the scientific enterprise as structured in cycles or phases (normal/revolutionary) characterized by methodological, perceptual, and semantic incommensurability. For the relevance of Kuhn's work for the early sociology of scientific knowledge, see Barnes 1982. For a conservative interpretation of the Kuhnian concept of paradigm ("used in such a way as to facilitate the separation of the description of scientists' social activity from the description of their cognitive activity," 466), as opposed to a radical interpretation ("taken to be a term which emphasizes the integration of, and the holistic nature of cognitive and social activity in science," 466), both influential for the early sociology of science, see Pinch 1997. For a critical review of Kuhn's later *Black-Body Theory and the Quantum Discontinuity 1894–1912* (1978), where "science is portrayed as a process much less susceptible to human or even social influence [since] nature is firmly in the driver's seat," see Pinch 1979b.

3. See in particular Merton 1938, 1973. Merton's functionalist approach addresses science as an institution, describing its forms of structural organization, its status-roles and values systems, its procedures for the allocation of resources and rewards, and its career paths. The new sociologists of science distanced themselves from the Mertonian approach as a sociology of scientists that keeps the production and validation of scientific knowledge "black-boxed."

4. For an introduction to science studies, see Hess 1997 and Sismondo 2010. For a recent critical overview on science studies as "naturalized philosophy," see Collin 2010, according to whom "adopting the perspective of naturalization means highlighting those aspects of Science Studies that constitute an attempt to replace, or at least to augment, traditional philosophical approaches to science with empirical ones, or to answer traditional philosophical questions by empirical means" (vii).

work,[5] Paul Feyerabend, Popper's student Imre Lakatos,[6] they were all avidly read. And with them Thomas Kuhn, who was a historian of science, with the furor produced by his book, *The Structure of Scientific Revolution*, and the debates with the philosophers. It was a very lively time, and these ideas were up in the air.

The sociology of scientific knowledge (as we call it, SSK) really got developed crucially by a group in Edinburgh, associated with the Edinburgh Science Studies Unit:[7] this was one of the first science studies programs in the

5. See Popper 1959, 1962, 1972. Harry Collins (1981c) acknowledges the relevance of Popper's falsificationism and "his stress on the temporary nature of contemporary knowledge" for the new sociology of scientific knowledge.

6. See Lakatos 1976, 1978, 1980. Lakatos underlines how Popper's falsificationism does not account for the survival of falsified theories (and for theories that arise "already falsified" by "known facts") as reported in the historical works of Thomas Kuhn. For Lakatos, scientists are involved in research programs articulated in a "hard core" of theoretical hypotheses that they try to protect from falsification through a "protective belt" of auxiliary hypotheses. Research programs can be "progressive" (able to extend the theory to include new facts and generate more accurate predictions) or "degenerating" (unable to generate a growth in knowledge), and it is up to the scientific community to discourage the second type, for example, through ostracism from scientific publications. In this regard, notwithstanding the existence of objective criteria of choice among different research programs in a particular historical moment, there would be no "epistemological algorithm" able to drive scientific enterprise alone: the appeal to the scientific community implicitly recognizes the relevance of a social dimension in the making of science.

7. The Edinburgh School in the sociology of scientific knowledge gathered around the Science Studies Unit, founded at the University of Edinburgh in 1966. David Bloor, one of its leading scholars, recalls those early years and the figure of David Edge (1932–2003), first director of the Science Studies Unit and future editor of leading journal *Social Studies of Science*:

The 1960s was a period of expansion in British universities. Governments believed in universities and were rightly confident in their ultimate value. Educational rather than purely utilitarian virtues mattered. One such educational value was the idea that science students should not be narrow specialists. The increasing influence of science and the increased status of scientists ought to be accompanied by a corresponding broadening in their education. C. P. Snow's "two cultures" had to be broken down. In response to these concerns, a number of British universities began to set up programmes of interdisciplinary study. ... Edinburgh's contribution to this general educational movement within science owed a great deal to the biologist C. H. Waddington. ... Waddington and his Committee set up the Science Studies Unit, and David was appointed to be its first Director. ... In 1966 there was no discipline called Science Studies, there was no template or clearly defined role. ... People were allowed to experiment and make up things as they went along. ... Creating a new department, and a new form of academic activity, from scratch was just the sort of thing to engage

world, and certainly in Britain. In particular, there were two very influential people. The first was David Bloor. He published an earlier piece, in 1973, "Wittgenstein and Mannheim on the Sociology of Mathematics,"[8] which was widely read. His book, *Knowledge and Social Imagery*,[9] which came out in 1976, was hugely influential. Bloor was one of the pioneers, together with his colleague, Barry Barnes,[10] who has worked with him at Edinburgh. Then

---

David [Edge]'s enthusiasms. ... David appointed a young and untried staff [including Bloor himself, Barry Barnes and the historian Steven Shapin], but one whose work had already been influenced by ... thinkers [like Kuhn, Lakatos, MacIntyre, Hesse]. ... Our collective approach gradually turned into a mode of analysis of science informed by philosophy, but fundamentally different from philosophy in its methods, being grounded in an empirical curiosity about the history and sociology of science. We came to realize that we were doing the sociology of knowledge. (Bloor 2003, 172–173)

For an account of those years at the University of Edinburgh, see Henry 2008. For a description of the courses taught at the Science Studies Unit, see Edge 1971 and Bloor 1975.
8. Bloor 1973. In this groundbreaking article, Bloor sketches the foundations of the Strong Programme for the sociology of knowledge, based on the requirements of causality, impartiality, reflexivity, and symmetry:

The first [requirement] is that the sociology of knowledge must locate causes of belief, that is, general laws relating beliefs to conditions which are necessary and sufficient to determine them. The second requirement is that no exception must be made for those beliefs held by the investigator who pursues the programme. ... The programme must be impartial with respect to truth and falsity. The next requirement is a corollary of this. The sociology of knowledge must explain its own emergence and conclusions: it must be reflexive. The fourth and final requirement is a refinement of the demand for impartiality. Not only must true and false beliefs be explained, but the same sort of causes must generate both classes of belief. This may be called the symmetry requirement. (Bloor 1973, 173–174)

While acknowledging its relevance for the Strong Programme, Bloor underlines how Mannheim's sociology of knowledge fails to achieve a full symmetry: mathematical truths, in particular, would not require/allow any sociological explanation, being self-evident as "truths as such." In this case, the sociology of knowledge would be but a "sociology of error," aiming only to explaining historical failures to recognize "truth." Drawing on the latest Wittgenstein (1956) and his "non-Realist Theory of the objectivity of mathematics ... [that refers to] Mathematics and Logic [as] collections of norms, [and to their] ontological status ... as that of an institution ... social in nature" (Bloor 1973, 189), Bloor concludes that "activities of calculation and inference are amenable to the same processes of investigation, and are illuminated by the same theories, as any other body of norms" (ibid.), opening the field for a full symmetry in the sociology of scientific knowledge.
9. Bloor 1976. The volume, in which the Strong Programme is fully articulated, is characterized by the philosophical sensibility and by the choice of historical case studies that will be recognized as typical of the Edinburgh school.
10. Barnes 1974, 1977; Barnes and Shapin 1979.

there was the guy I ended up working with, Harry Collins, who was at Bath University[11] and started to publish a little later. I think my recollection is that Collins's early articles were maybe from '74, '75. Studies of TEA lasers[12] and gravity waves:[13] a very influential early work.

I think one difference, in the different approaches, was that Bloor was primarily a philosopher. He was trained as a psychologist, but his argumentation was largely philosophical. So in that book, *Knowledge and Social Imagery*, he engaged with philosophers and tried to reason what a

11. The Bath School, the main affiliates of which are Harry Collins and Trevor Pinch, is known for its Empirical Programme of Relativism (Collins 1981b, 1981c): strongly grounded in empirical research, the program shares with Edinburgh's Strong Programme the principles of symmetry of explanations and impartiality toward success or failure of a theory, but it dismisses the call for causality and suspends the call for reflexivity. As will be discussed in more detail later in the interview (see secs. 1.2–1.4), the program is based on three steps: illustrate the "interpretative flexibility" of experimental data, clarify how a "closure" of such flexibility is reached, and link the closure mechanisms to the wider social context (Collins 1981a). These three steps will be adapted by the SCOT model (see secs. 3.1–3.2) to the social construction of technological artifacts. For a list of empirical studies acknowledged by Collins in the early 1980s as "within the relativist programme" (including works by Dean, Barnes, Shapin, Harvey, MacKenzie, Wynne, Pickering, and Pinch, among others), see Collins 1981a, 1981c.
12. Collins 1974; Collins and Harrison 1975. In these studies, Collins shows how a large part of the scientists' expertise and knowledge is actually tacit (in particular, the expertise needed to set up a working TEA laser system), and, as such, impossible to formalize and transmit by academic papers.
13. Collins 1975, 1981b. Drawing on fieldwork done in 1972 and 1975, these two papers focus on the detection of high fluxes of gravitational radiation claimed by Joseph Weber and on the unsuccessful attempts to replicate the original experiment by different research teams. In the first paper Collins shows how, the actual existence of high fluxes of gravitational radiation being controversial, the scientific community lacked common criteria to define a properly working experimental set. As a consequence, mismatching results triggered a potentially never-ending controversy on which experiment should be considered valid: a phenomenon Collins calls "experimenter's regress." The second paper shows how, by 1975, the controversy over the existence of high fluxes of gravity waves was over: as Collins points out, "the existence of (hf) gravity waves ... [was by then] literally incredible. My claim is ... that their demise was a social (and political) process" (37).

sociology of knowledge—that wasn't just a sociology of error—would look like. The key move toward what became known as the Edinburgh Strong Programme was to avoid what Bloor used to call "epidemiological approaches to knowledge," where you just would apply the social to the error, to explain mistakes socially, as if knowledge got diseased by mistakes. True knowledge—so-called true knowledge—and false knowledge were both to be explained by the same type of sociological explanation. This is the need for symmetry, where one does not have one explanation for "true knowledge" and a different sort of explanation for "false knowledge"—say, rationality for true knowledge and then sociological or psychological error for false: that would be an asymmetrical explanation. He was very committed to this, and he outlined this in what became known as the four tenets of the Strong Programme in the sociology of scientific knowledge. The four tenets were symmetry, impartiality, causality, and reflexivity. That was incredibly influential.

But there were other people doing similar things,[14] as you might expect. You know, historical hindsight makes it look like it was just, you know, Barnes, Bloor, and Collins, but there were others. I worked with a guy in Manchester, called Richard Whitley, who had written a very early important article in this period as well, called "Black Boxism and the Sociology of Science,"[15] which was an attack on the Mertonian approach. The predominant approach in the sociology of science of that time was that of Robert Merton; Whitley's contribution was to famously claim that Merton treated, I remember, scientific knowledge like the scientist could be producing pies as much as scientific knowledge: it made no difference. The actual content of the science did not seem to figure into the Mertonian approach.

14. Harry Collins (1983c) includes "six independent contributors" in the initial phase of the sociology of scientific knowledge: Michael Mulkay (1969), David Bloor (1973), Richard Whitley (1972), R. G. A. Dolby (1974), Barry Barnes (Barnes and Dolby 1970), and himself (1974).
15. R. Whitley 1972. "Whitley demanded that scientific knowledge be opened up to examination. In the current sociology of science, he argued, production of scientific knowledge is treated as a 'black-box,' of which only the inputs and outputs can be studied. He suggested that a sociology of scientific knowledge will have to open up the box" (Collins 1983c, 269).

Then, of course, there were other attacks on Merton: Barry Barnes, whom I have just mentioned, wrote an article with a guy called Dolby on the Mertonian ethos;[16] Michael Mulkay wrote an important article about the norms of science.[17] So these attacks on Merton saw a new approach developing. And there were other Europeans, later in the mix: Bruno Latour,

16. Barnes and Dolby 1970. In Merton's functionalist approach, "science is treated as a social institution with an ethos of norms and values" (3): universalism, communism, disinterestedness, and organized skepticism (Merton 1957). In this article, Barnes and Dolby "aim ... to show that Merton has failed to identify a constant, specific, overriding normative structure within which this activity occurs" (7), having ignored any distinction among professed norms and statistical norms, together with any form of historical transformation of scientific norms. For Barnes and Dolby, "within the highly differentiated societies of today social order can be maintained independently of a total normative consensus. ... Those groups of scientists showing the greatest degree of consensus are Kuhn's paradigm-sharing communities. The cohesion, solidarity and commitment within these stem from the technical norms of the paradigms, not from an overall scientific 'ethos'" (23), while "normative dissent clearly plays a vital role in science" (24).

17. Mulkay 1976. Like Barnes and Dolby, Mulkay rejects Merton's description of a set of institutionalized norms as a crucial feature of the modern scientific community, "accounting for ... [its] rapid accumulation of reliable knowledge" (638). In front of "an exactly opposite set of ... [counter-norms] essential to the furtherance of science" (639) empirically described by Mitroff (1974), "the original set of functionalist norms" would be better understood as an ideology, a discursive resource strategically advocated by scientists when "portraying and justifying their actions to lay audiences" (646),

vocabularies of justification, which are used to evaluate, justify and describe the professional actions of scientists, but which are not institutionalised within the scientific community in such a way that general conformity is maintained. ... The leaders of academic science in Britain and, more clearly, the United States have drawn selectively on these vocabularies in order to depict science in a way which justified their claim for a special political status; ... the biased image of science which they have vigorously proclaimed seems to have been widely accepted, not only among the public at large, but at least in part within official circles. (653–654)

SSK should aim at describing how different vocabularies are used within the scientific community and at accounting for the widespread acceptance of an ideological description of science.

Karin Knorr-Cetina,[18] and Peter Weingart,[19] whose work was influential in the early days. And there were also Michael Lynch and Sharon Traweek,[20] who both started to do laboratory studies about the same time as Latour and Knorr-Cetina, but only published much later. So it was a mixture of different approaches.[21]

18. In the early years of SSK, Knorr-Cetina (1983) distinguished two microsociological approaches: "The first approach [proposed by the Bath School] focuses on scientific controversies as a strategic anchoring point for the study of consensus formation, that is, of the mechanisms by which knowledge claims come to be accepted as true. ... The second approach has chosen direct observation of the actual site of scientific work (frequently the scientific laboratory) in order to examine how objects of knowledge are constituted in science" (117). In her early ethnographic work, Knorr-Cetina observed for a year (from October 1976) a research group working on plant proteins in a government-financed research center in Berkeley, California (Knorr-Cetina 1981a; see also Knorr 1977, 1979; Knorr-Cetina 1981b), while Latour observed for two years (from 1975) the work of the research group in endocrinology led by Roger Guillemin at the Salk Institute, La Jolla, California (Latour and Woolgar 1979). For a critical overview of early laboratory studies, see Woolgar 1982, where these two groundbreaking studies are contrasted as examples of instrumental vs. reflexive ethnographies: in the first case, "the ethnographer is concerned with finding things to be other than you supposed they were ... [and] to argue that science is 'in fact' quite an ordinary enterprise ... [and that] scientific work can be shown to be similar to non-science" (486–487), while "by contrast, reflexive ethnographies of scientific practice are not merely intended to produce news about what goes on in laboratories. ... The ethnographic study of the laboratory is an occasion for investigating scientific practices for what this can tell us about practical reasoning in general: ... an ethnography of scientific practice should be a study in a laboratory, not just a study of a laboratory" (487).
19. Weingart 1974. "Weingart ... has specified the different elements of a paradigm and outlines their hierarchical relationship" (Pinch 1997, 401): this contribution is listed by Trevor Pinch among the conservative interpretations of Kuhn that "undermine the power of the paradigm concept as a means of integrating the social and cognitive aspects of science" (ibid.). See also Mendelsohn, Weingart, and Whitley 1977.
20. Michael Lynch (1985a) applied conversational analysis to "shop talk" in a university neuroanatomy laboratory in California in 1975–1976. For a review, see Pinch 1987b. Sharon Traweek started her five-year-long ethnographic fieldwork in spring 1976, "at three national laboratories ...: National Laboratory for High Energy Physics (Ko-Enerugie butsurigaku Kenkyusho, or KEK) at Tsukuba, Japan, Stanford Linear Accelerator (SLAC) near San Francisco, and Fermi National Accelerator Laboratory (Fermilab) near Chicago" (Traweek 1988, 11).
21. For a general overview of the main approaches in the early 1980s, see Knorr-Cetina and Mulkay 1983.

ST   Why did it become so urgent to try to open the "black box" in those years?

TP   That is an interesting question: why were people so obsessed with opening the black box? Actually, I do not know, but it seemed the right thing to do. Many of the people coming into this had some background in the sciences. I know Collins had some background in physics; I had an undergraduate degree in physics; later Pickering came in the field, he had a PhD in physics. Barnes had some technical background. Many of these people were familiar with science, and I think they just found Merton's approach dissatisfying. They knew enough about science to know that there had to be more to say—more interesting than just the normative or institutional structures of science, which Merton was doing, which was of course a fine sociological approach. But if you have done some science, you know for example about these entities called neutrinos. For me, it's dissatisfactory just to talk about the physics community or the career structure of physicists: we wanted, somehow, to be more evocative. And the philosophers of science *were* talking about the particularities of knowledge; so you read Karl Popper, Imre Lakatos, or Paul Feyerabend and their case studies. You know, Feyerabend had this great study on Galileo. Popper had case studies on the reception of Einstein's theory of relativity. Or Kuhn, who was incredibly influential: his stuff was grounded in detailed knowledge of what was happening in the sciences. I think Kuhn, by the way, was a sort of a missing link in all of this. Because everybody read *The Structure of Scientific Revolution*: it was like a framing book. In fact, Barnes went on to write a book about T. S. Kuhn and social sciences,[22] which is all about Kuhn's influence. It was just like the natural starting point. So Kuhn had opened the black box: he had opened it historically. Having read Kuhn, you immediately would think: Well, can we do the same sort of work contemporaneously? It was a kind of a natural following-on from these historical studies, and since we all had some technical expertise in the sciences, it seemed that that was the starting point to say more.

However, Collins—and, like him, many people—just stumbled into this thing: in hindsight, you can seem to trace rational paths, but it was not like that. I know Collins, who was my closest associate, the coauthor of the *Golem* books,[23] my mentor and PhD supervisor, he stepped into this

22. Barnes 1982.
23. See section 2.1.

field because he had done a master's degree at the University of Essex sociology department. In Britain at that time, Essex was a cutting-edge sociology department—Colin Bell[24] and Howard Newby[25] were there. Barry Barnes was also at Essex at one point. And this was an interesting time in Britain because all you needed to get a job as a professor at a university back then—we say professor in the American sense, in Britain it would be what we call a lecturing position, but your first proper academic job—all you needed was a master's degree. Sociology in the United Kingdom was expanding. Collins couldn't get a job immediately so he started a PhD at the University of Bath and after two years he got a grant from the U.K. Social Science Research Council (SSRC) and obtained a lecturing position at Bath. At Essex he had started working on a pretty conventional project, some social networking-type problem, more akin to the old Mertonian approach, which was done by people like Nick Mullins and Diana Crane: they were mapping out the invisible colleges of science,[26] you know. So, network analysis. Collins started off, I remember, doing this kind of network analysis—I am not sure he even knew formally about the work on networks in the United States, as Collins was pretty much self-taught—before he stumbled into this idea, when he was following a network of

24. The British sociologist Colin Bell (1942–2003), a scholar in class and race relations, community studies, and methodology of the social sciences, lectured at Essex University from 1968 to 1975.
25. Howard Newby lectured at the University of Essex until 1979, when he was appointed Professor of Sociology and Rural Sociology at the University of Wisconsin. He returned at Essex as Professor of Sociology in 1983.
26. "Invisible Colleges" is the term used by Diana Crane to refer to recognizable social networks of scientists that shape communication patterns within the broader scientific community and, consequently, the diffusion of knowledge. Both Mullins's (1972b, 1973; see also Mullins 1972a, Groeneveld, Koller, and Mullins 1975, Mullins, Hargens, Hetch, and Kick 1977) and Crane's (1969; see also 1965, 1972) contributions are listed by Pinch (1997) among the conservative interpretations of the concept of paradigm:

Kuhn's association of paradigms with specific social groups ensures that they can be located by reference to the social characteristics which define such groups alone. In other words, paradigms can be located by the tools of sociometric analysis. For instance, the communication patterns and citation behavior of a group of scientists could be used to identify the existence of a paradigm. Within this interpretation, paradigm is to be treated as another term (like "invisible college," "specialty" and "discipline") and descriptive of social networks in science. This use of "paradigm" entails a firm dividing line between the cognitive and social elements of science. (466–467)

scientists building a new kind of laser.[27] And he suddenly realized there was an issue: what was being passed along the network? A scientist builds a laser in Canada, where it started, and a group in the States cannot replicate building that laser. So what knowledge is being passed on, and what failed to be passed on? And he was thinking about it in terms of social networks and then quickly realized: "Oh! To do this properly, one has to get into the content of the knowledge, the practices the scientists are passing on or failing to pass on." Bingo! Suddenly he had the whole idea of tacit knowledge at the core of the sociology of scientific knowledge. So he stumbled into it by initially tracing out networks and suddenly realizing the need for something new. I think Barnes and Bloor, because they did not do so many empirical studies, thought about it more conceptually. Bloor, in particular, was more interested in the philosophy of mathematics: what would be needed to do a proper philosophy of mathematics? And he'd read Wittgenstein, Frege … and he was involved in some more philosophical debates. Barnes was more of a sociologist, but not an empirical sociologist, even if he did do some empirical work. So it was a complicated mixture of things. If you are talking about opening the black box, it was definitely felt by these groups of people, for these reasons.

1.2   The Bath School: Scientific Controversies and Tacit Knowledge

ST   So, what we had at that time was a shared intellectual background—the work of historians, like Kuhn, and philosophers—that was pushing toward a deconstruction of the textbook view of science. Then we had a group of people with a very peculiar expertise, like you: people who somehow had firsthand experience of science, who perceived a mismatch between the emerging idea of science and its ordinary representation. These scholars could not find in the mainstream Mertonian sociology of science the right instruments to fill this gap. From this point of view, the key challenge was somehow methodological: the problem was how to address the way science was—is—actually being done, science as practice …

TP   Exactly. We had no clue, back then, as to how to do this work methodologically. We all stumbled into this somehow by accident.

27. Collins 1974; Collins and Harrison 1975.

Collins stumbled on the idea of interviewing a "core set"[28] of the scientists involved in an interesting area of development of new knowledge by accident. The same for me: I did my first research in my master's degree at Manchester University on the scientific controversy over hidden variables in quantum mechanics.[29] And, I remember, it was not like I had a methodology. The guy I was working with, Richard Whitley, had six different projects. We were all students of his, and he said: "Here are six projects that I'm interested in." I cannot remember any of them—they were all kinds of official projects. They were also incredibly boring to me. Then he mentioned David Bohm and the foundations of quantum mechanics, and I thought: Yes! That's me! I want to do that one!—before anyone else could claim it. Because I came out of physics, I was dissatisfied with physics, and as I was always interested in this wave-particle duality issue, suddenly I thought: Why can't I study this for a sociological project? Great! I was actually studying what turned out to be a controversy, but I didn't know what a controversy was back then or how to study one. This was a historical one; the main issues were in the 1950s. All I knew was that I was working with a sociologist, Richard Whitley, who told me, "You've got to go and interview Bohm." He forced me to go and interview the main scientist. So I kind of stumbled into the whole idea of studying scientific controversies. And it was only later, when I worked with Collins, that we came up with some kind of controversy methodology, and we formalized it. But back then, you know, you just stumbled into a circle of scientific controversy. Isn't that interesting?

ST   Indeed! In fact, your approach to this controversy in quantum physics involving Bohm was quite different from what you did later. In this first work, you focused mainly on what makes communication difficult across different paradigms.

28. "The set of allies and enemies in the core of a controversy are not necessarily bound to each other by social ties or membership of common institutions. Some members of this set may be intent on destroying an interpretation of the universe upon which others have staked their careers, their academic credibility and perhaps their whole social identity. ... This set of persons does not necessarily act like a 'group.' They are bound only by their close, if differing, interests in the controversy's outcome. I refer to such a set of allies and enemies as a 'core set'" (Collins 1985, 142–143).
29. Pinch 1977.

TP Yeah, it was grounded more in Kuhn,[30] and also, this was before I met Collins. Whitley was very into reading this economist called Nicholas Georgescu-Roegen, who had this idea of "arithmomorphism"[31] in science: there was a tension in different areas of science as to how to do mathematical work. Whitley wanted me to use that approach, so that article has quite a bit of that stuff about communication issues around mathematics. I think I would have rather done it in a different way. Later on, I actually did write that article in a different way,[32] in the early publications of a British journal, *Physics Education*, on controversies, and then I wrote about Bohm. I think that was a typical scientific controversy,[33] like the ones we were studying in

30. "One of the more interesting problems to arise from the Kuhnian analysis of science is the problem of communication between scientists with differing cognitive commitments. ... My concern is to understand communication problems which arise in the actual development and practice of science. ... This paper is intended to explore a communication breakdown over a specific cognitive object, namely a proof that a certain type of theory in quantum mechanics is impossible" (Pinch 1977, 171–172).

31. Pinch 1977, 202:

Arithmomorphic concepts are concepts that are amenable to logical (in the narrow Aristotelian sense) sifting. Georgescu-Rogen argues that logic can handle a restricted class of concepts, that is arithmomorphic concepts. This is because "every one of them is as discretely distinct as a single number in relation to the infinity of all others." ... He is particularly interested in distinguishing such concepts from dialectical concepts, that is concepts which emphasize forms and qualities. Dialectical concepts are specifically defined as concepts which violate the Principle of Contradiction "B cannot be both A and not-A." He argues that dialectical concepts are as relevant to science as arithmomorphic concepts and that the barring of dialectical concepts from science leads to a situation of "empty axiomatisation" and "arithmomania."

See Georgescu-Roegen 1971.

32. See Pinch 1979c.

33. Pinch 1979c, 48:

Scientific knowledge is a social product constructed, and indeed fought for, by scientists in particular social and historical settings. In this view it is at times of scientific controversy, when all or part of such knowledge comes under challenge, that the social dimensions of science become clearest. Just as natural scientists often learn most about the system under study when it is experiencing its greatest perturbation or stress, so too can the student of the scientific enterprise find the stress produced by a scientific controversy most rewarding for revealing the social processes of science. The hidden-variables controversy, which raged so vociferously in the 1950s and early 1960s following the challenge posed by hidden-variables theories to the orthodox version of quantum theory, is a good example of a controversy which was obviously permeated by social influences.

Bath later. So I reinterpreted it as a standard scientific controversy. But you are right: at that time my approach was somehow different. I loved doing interviews. We had not worked out why we were studying it that way.

ST So after this first work, you (and, of course, your group) seemed to have finally found a way to open the black box: the key was a focus on controversies and on the actual practices and expertise of the scientists involved.

TP Exactly.

ST Can you explain how you used controversies as the key to open the black box?

TP Well, one of the things we were stumbling around by then was "What exactly is the nature of scientific expertise?" It was not really very clear back then. So the Bohm project for me was largely an interview- and textual-based project. I looked at texts in the history of science. More precisely, not texts *on* the history of science: texts in physics, examined through the lens of the history of science. I was finding old physics articles about this controversy; I had interviews as well, but they were kind of historical interviews.

The spur to look at something contemporary and look at practices actually came from Collins, who, after his study of lasers, hit upon the idea that it would be interesting to look at fringe science, paranormal science. He got a Social Science Research Council grant to pursue this. He was interested in that because—I don't know, maybe he actually had an interest in fringe science, at one point, but mainly because stuff at the margins of science will tell you more about what was going on in science, he thought. If you can look at how science rejects stuff off the edges, you can find out more about how science works.[34] We were interested in these problems of

---

34. Pinch 2014:

Many of us who worked in the early sociology of scientific knowledge (SSK) cut our teeth on fringe science (e.g., Shapin 1975; Wallis 1979; Collins and Pinch 1982). It was important in arguments with philosophers of science to show that their demarcation criteria for separating the wheat of science from the chaff of error were untenable. Later, what Collins termed the "hard case" for SSK (physics) took priority over the "marginal case." The SSK research project was to make progress against the head winds coming from the philosophy of science by tackling between the most credible science, such as physics, (which was hard to show as social) and the fringe sciences, such as parapsychology, which philosophers considered could only be social because they were in error. With the battle against the philosophers of science won by the mid-80s, the issue of demarcation was largely left behind.

demarcation[35] of real science from pseudoscience. He thought of it almost as a methodological principle, and he was studying this contemporaneously. He had heard about this psychic *wunderkind* called Uri Geller, who amongst others things claimed to be able to bend metal by the power of mind alone.[36] And he quickly found that, in his own university, there was a physicist who was interested in studying Uri Geller. So Collins, as a sociologist who was familiar with the methodology of participant observation, thought it was a great opportunity: there is a scientist there actually researching this, he thought, why don't I just team up with him, and study him at the same time as I study the kids that are claiming to have the same abilities as Uri Geller?[37]

And that is where he developed his whole methodology; he actually calls it "participant comprehension" rather than "participant observation." He later wrote an important article on this distinction[38]—because often,

35. See Pinch 1979a:

this paper examines the problem of demarcating genuine science from "pseudoscience." It is shown that it is possible to turn the demarcation arguments which have been used against "pseudo-sciences," such as parapsychology, against the fraud hypothesis—which is the principal normal counter-explanation for the parapsychological evidence. It is argued that the fraud hypothesis fails to be scientific on the grounds of replication, metaphysical bias, falsifiability and lack of theory. Since fraud is accepted and parapsychology rejected, the role of demarcation criteria in determining acceptable science is challenged. An alternative account of their role is presented. It is argued that the rejection of parapsychology rests on cultural differences which demarcation criteria serve to legitimate. (48)

See also Pinch and Collins 1979, 1984; Collins and Pinch 1979, 1982.

36. Uri Geller (Tel Aviv, 1946) is a media celebrity, famous for his claimed psychic abilities, especially metal bending, as shown worldwide in many stage and TV shows since the late 1960s.

37. In 1975, Harry Collins and Brian Pamplin were conducting experimental research at Bath University with six young kids (from eight to thirteen years old) who claimed the ability to bend objects in the way shown at that time by Uri Geller on British Television (Pamplin and Collins 1975).

38. See Collins 1979, 1983b, 1984. For Collins, "participant observation" can be divided into "unobtrusive observation" and "participant comprehension." "Unobtrusive observation," whose "archetypal study" is represented by Festinger's *When Prophecy Fails* (Festinger, Riecken, and Schachter 1956), is informed by a positivistic stance: "the essence of the method is that the investigators want to observe the actions of others while disturbing those actions as little as possible. The investigator wants to disturb the situation as little as possible because he believes (as a good positivist) that it is his (read 'his or her') job to discover 'what is really going on

in participant observation, people think the goal of sociology is somehow not to immerse yourself too much, to worry about not going native and keeping your objectivity, and not getting too immersed. Well, Collins is precisely the opposite. You welcome participation and comprehension; you immerse yourself in this paradigm. The methodological idea is: from your phenomenological immersion in the same practices of the scientist, you actually understand what they are doing. It is very different from the way social psychologists like Festinger[39] talk about participant observation, where somehow you are constantly worried about being biased or getting too involved. This was "throw yourself in." Collins was very influenced by the later ideas of Ludwig Wittgenstein on how language and the world are coconstructed within a "form of life."

So the practices just seemed to be an important thing to look at, at that time, because you were following the scientists doing experiments. Kuhn had already pointed to the importance of practices in science; he talked about the training of scientists: the importance in the notion of paradigm, not only of developing theories and metaphysics, but also lab practices. So the whole idea of a paradigm was caught up in this. And this is all in this book, *Frames of Meaning*,[40] actually, that Collins and I wrote. The whole idea of a paradigm is a notion that combines praxis with theory. That is the key to understanding the term *paradigm*, because it is very different for some people. Some people think of paradigms like worldviews, therefore only theory. But for us it was a term that encapsulated practice and theory together. The example we had in mind—this is another influential book—was a Wittgensteinian book by a philosopher called Peter Winch.

---

there' and, hence, that 'observer effects' are bound to be distorting. ... The distinction between observer and observed is absolute" (Collins 1984, 56–58). In contrast, "participant comprehension," as practiced by Collins and Pinch in their research on fringe science (Collins and Pinch 1982), is ascribable to the interpretive tradition:

In participant comprehension, participation is not an unfortunate necessity ...: it is not risky or full of pitfalls. It is central, irreplaceable and, indeed, the essence of the method. In participant comprehension, the participant does not seek to minimize interaction with the group under investigation, but to maximize it. Native incompetence is not a technical problem to be overcome at the outset ... but rather the development of native competence may be the end point of participant comprehension. ... Comprehension will have been achieved when what once seemed irritating and incoherent comes to follow naturally. (Collins and Pinch 1982, 61)

39. Festinger, Riecken, and Schachter 1956.
40. Collins and Pinch 1982.

Collins was very influenced by this book: Peter Winch, *The Idea of a Social Science*.[41] There, Winch has a similar example to one of Kuhn's, which is the germ theory of disease. He is considering what happens when the germ theory of disease takes off.[42] You can look at it through a Kuhnian's eye because you not only get a new theory, you have new practices. So surgeons, once they see germs for the first time, they have to get rid of them, so they start to scrub up, they cover their body in drapes, they wear masks. And this is not just an idea: it changes the way that medicine is performed. We were very influenced by that example because it showed that practices and ideas are tied in together. So a sociology of scientific knowledge had to look at practices as well as at the ideas. And that is what we did.

There is often a big misconception in the field. Some people say, "Oh, the focus on practice came later with a book by Andrew Pickering, called *Science as Practice and Culture*,"[43] or "It came from laboratory studies, what

41. Winch 1958.

42. Winch 1958, 120–121:

Now compare with … [the] discovery of a new germ the impact made by … the first introduction of the concept of a germ into the language of medicine. This was a much more radically new departure, involving not merely a new factual discovery within an existing way of looking at things, but a completely new way of looking at the whole problem of the causation of diseases, the adoption of new diagnostic techniques, the asking of new kinds of questions about illnesses, and so on. In short it involved the adoption of new ways of doing things by people involved, in one way or another, in medical practice. An account of the way in which social relations in the medical profession had been influenced by this new concept would include an account of what that concept was. Conversely, the concept itself is unintelligible apart from its relation to medical practice.

43. Pickering 1992b. In the introduction of the volume, Pickering remarks how "taken seriously as an image of practice and culture rather than as an aid to thinking about knowledge, SSK's account is thin, idealized, and reductive" (Pickering 1992a, 5); and how, while the attempt to "enrich [the understanding of scientific practices] through empirical study … has been a main line of development within SSK" (see Gooding, Pinch, and Schaffer 1989), the "analytic repertoires developed in the service of a problematic of knowledge [cannot] serve as the primary basis for understandings of practice" (Pickering 1992a, 5–6). Pickering will try to enrich these repertoires through the conceptual metaphor of the "mangle" (Pickering 1995). As for Pinch's commentary (1999, 139–140), "the mangle in its elementary form is the dialectic between human and material agency. In the course of practice (in the sense of activities occurring in real time), such as building a new scientific instrument …, or a new technology …, human agency and material agency get intertwined. Human agency, in the shape of goals, encounters resistance from the material world, in the shape of non-human agency. In accommodating to this resistance during

Bruno Latour and Knorr-Cetina were doing." But no, that's not true. As the group in Bath associated with Collins, we were studying how experiments were done in practice, in the laboratory, at our own laboratory, watching the scientists. So we were very interested in practices. In fact two whole chapters of that book, *Frames of Meaning*, are about the experimental practice in doing paranormal science. It is a study of experimental practices; it's a participant observation study in a lab. The idea that somehow we were just doing controversy studies is wrong. We actually spent quite a bit of time in the lab, doing laboratory studies. That book is less well known than it should be, because we picked a case of a revolution in science, but since we picked a failed revolution (because paranormal stuff never took off), the book is kind of forgotten. The book did not get a wide recognition since the book is on parapsychology, but it really was also about lab practices.

ST This is actually interesting: the failure of the object of your study, parapsychology, somehow determined the fact that the book is underknown and often forgotten …

TP It's true! The two things should not be connected, but they are. Parascience and everyone dealing with that topic experienced similar problems. David Hess, for example: he wrote an article many years later.[44] He also studied para-science, and everyone studying paranormal, studying this weird stuff, has this stigma attached to them or to their work. What is important, and Collins has always made this point to me, is never to be caught only doing fringe stuff. This fringe stuff is fun, and you learn a lot,

---

real-time practice, human agency and material agency get 'mangled.'" For Pickering, this approach would overcome the boundaries of SSK, "with its roots firmly set in humanism." While, in fact, the mangle approach "strive[s] to establish a space where human actors are still there, but are inextricably entangled with the non-human," within SSK, "non-human agency [would be] always translated … as 'accounts of non-human agency', thus leaving humans in ascendancy" (Pinch 1999, 141). As we will see, from the epistemological and methodological problem of nonhuman agency will stem one of the longer-lasting quarrels within science and technology studies (STS) (see section 3.5). For Pinch, while "SSK analysts … are prepared to grant that … actors' identities can in part be constituted by objects," the "inextricable entanglement of human and non-human" doesn't imply that "we have to give up on social explanation; the social explanations just have to be more sophisticated" (Pinch 1999, 145) than those of early SSK.

44. See Hess 1996. See also Hess 1993.

but you always have to "tack": he called it "tacking," like a sailing boat tacks backwards and forwards. You have to tack back to the harder core of sciences, where people cannot dismiss you for only working on that, for only indulging in fringe science.[45] That word, "indulging," is a bad word for what we were doing.

ST Clearly it is. So, let's take Collins's advice and let us go back a bit to the role played in your approach by this peculiar attention to practices, and the role it played in your effort to open the black box. I would love to know, possibly on the basis of some examples from your case studies, how scientific knowledge is actually linked to practices and to expertise and not only to a formal network of scientific concepts.

TP Well, the earliest, best example of this was not in my own work but in Collins's work on Transversely Excited Atmospheric (TEA) lasers,[46] that early study he did where I was telling you he was trying to trace why scientists failed to repeat a particular experiment. Why did they find it so hard? You have a laser built in one lab, why was it so hard for these other guys to build the laser and get it to work? And as he followed this group backwards and forwards and interviewed them, he realized that the only way they could get their laser to work was when they actually visited the site of the original experiment. He thought: What is going on? Why do they need to visit the site of this original experiment? And that is when he came up with the idea that a key part of science is practice: that it is very hard to articulate what you're doing.

He had not read Michael Polanyi, the British physical chemist who wrote about this in the 1950s, but Polanyi had a term for this: "personal knowledge" or "tacit knowledge."[47] So Collins wrote this

45. Pinch "indulged" in paranormal and fringe science in Pinch 1979a, 1984, 1987a. Together with Harry Collins, in Collins and Pinch 1979, 1982; Pinch and Collins 1979, 1984.
46. Collins 1974; Collins and Harrison 1975.
47. Polanyi 1958, 1966. As Polanyi summarizes:

I regard knowing as an active comprehension of the things known, an action that requires skill. Skillful knowing and doing is performed by subordinating a set of particulars, as clues or tools, to the shaping of a skillful achievement, whether practical or theoretical. We may then be said to become "subsidiarily aware" of these particulars within our "focal awareness" of the coherent entity that we achieve. Clues and tools are things used as such and not observed in themselves. They are made to function as extensions of our bodily equipment and this involves a certain change of our own being. Acts of comprehension are to this extent irreversible, and also

article[48] where he uses the term "tacit knowledge," knowledge that cannot be explicated. He went back to Polanyi's original example, which was riding a bike,[49] the moment we have a notion on how to ride a bike. We push along our kids: one moment they can't ride the bike, the next moment they can. They clearly learn something new, something that's being passed on, but what has been passed on has not been explicated. This is of key relevance: knowledge and practices, and craft practices, seemed very important. So that was one of the first studies. By the way, there was another early book that people read at the time, which also emphasized the craft side of knowledge, by a historian of science called Jerry Ravetz.[50] He wrote a book on the social processes of science. He had the notion of craft practices as well, the practices that scientists had, so that was a key thing: starting to look at the actual practices of the scientists and the idea of tacit knowledge. It was one of the earliest ones.

---

non-critical. For we cannot possess any fixed framework within which the re-shaping of our hitherto fixed framework could be critically tested. Such is the personal participation of the knower in all acts of understanding. But this does not make our understanding subjective. Comprehension is neither an arbitrary act nor a passive experience, but a responsible act claiming universal validity. Such knowing is indeed objective in the sense of establishing contact with a hidden reality; a contact that is defined as the condition for anticipating an indeterminate range of yet unknown (and perhaps yet inconceivable) true implications. It seems reasonable to describe this fusion of the personal and the objective as Personal Knowledge. (Polanyi 1958, iv)

48. Collins 1974.

49. Polanyi 1958, 51–52:

From my interrogations of physicists, engineers and bicycle manufacturers, I have come to the conclusion that the principle by which the cyclist keeps his balance is not generally known. The rule observed by the cyclist is this. When he starts falling to the right he turns the handlebars to the right, so that the course of the bicycle is deflected along a curve towards the right. This results in a centrifugal force pushing the cyclist to the left and offsets the gravitational force dragging him down to the right. This manoeuvre presently throws the cyclist out of balance to the left, which he counteracts by turning the handlebars to the left; and so he continues to keep himself in balance by winding along a series of appropriate curvatures. A simple analysis shows that for a given angle of unbalance the curvature of each winding is inversely proportional to the square of the speed at which the cyclist is proceeding. But does this tell us exactly how to ride a bicycle? No. You obviously cannot adjust the curvature of your bicycle's path in proportion to the ratio of your unbalance over the square of your speed; and if you could you would fall off the machine, for there are a number of other factors to be taken into account in practice which are left out in the formulation of this rule. Rules of art can be useful, but they do not determine the practice of an art; they are maxims, which can serve as a guide to an art only if they can be integrated into the practical knowledge of the art. They cannot replace this knowledge.

50. Ravetz 1971.

Then, for our group at Bath, the next important breakthrough Collins made came with another paper he wrote. See, now everyone thinks about books, but back then it was about papers; it was a sociological field where, actually, journal papers made it more than books, and that was kind of interesting. I'll tell you: when Latour and Woolgar wrote *Laboratory Life*,[51] nobody thought it would make a big difference because we all thought it was going to be an article field, not a book field. Then suddenly that book they wrote was getting an enormous mass of attention. "Oh my God, what we need in our field is to write books!" But no one knew that back then. Collins did his work in articles, and so he wrote a second article after that one on the TEA lasers. One of the cases he picked was an experiment on gravitational radiation by this Maryland physicist named Joseph Weber. Weber had found too much gravitational radiation, and other groups were having problems replicating his experiment.[52] Collins was looking at it, as a study similar to his TEA lasers study. So he was following the scientists in the network, and he tells a story of a discovery moment, a "road to Damascus" discovery; he was driving across the desert, in between inter-viewing these scientists, and he suddenly had this conceptual leap, where he saw that the issue here was that you didn't know if the knowledge that was being passed on even *existed* as valid knowledge. That was the very issue. In the TEA laser case, there was a clear measure if a laser would work: you could vaporize a block of concrete. In this case it was very different because you had groups building these massive bars that were shaking, basically for the detection of gravitational radiations, but you didn't know if gravitational waves actually existed—if they were supposed to shake or not. That is a great insight! It seems stupid, but that was something like a discovery. Wow! And this is what led him to this whole idea of "experi-menter's regress."

1.3   The Bath School: The Experimenter's Regress and the Externality of Observation

TP   So, the experimenter's regress: the very idea about who would count as the skillful, competent experimenter would depend on whether the claims

---

51. Latour and Woolgar 1979.
52. Collins 1975. See also Collins 1981b.

existed or not. In a controversy, you get caught in this loop[53] because, for one group deemed competent, the phenomenon exists. Then another group deemed competent does not find it: for them, it won't exist. So which experiment is a competent experiment?

ST I see: that would be decided by the existence of the claim, but the existence of the claim is the very object of the controversy.

TP Exactly. You're caught in this regress because you have to decide to build another experiment to decide if the phenomenon exists or not. It was a very simple, beautiful point, and he just stumbled into it, and he wrote this paper, "The Seven Sexes,"[54] which contains the key point. This was a major discovery in the field for us. I think that the more exact description would be that the experimenter's regress applies to cases of controversy at the research frontiers of science, where the outcome, either experimental or theoretical, is contested. So nobody knows for sure whether a new physical phenomenon (like, more recently, cold fusion, or high fluxes of gravitational radiation back then) exists or not. There is a controversy over their existence. You have experiments pro and con gravitational radiation, and you cannot tell who the skillful experimenters are, who has the competence. Are the ones who are finding gravity waves the skilled ones? That depends on whether gravity waves *are* there to be found, in which case they are the skilled ones. Once you get caught in this regress, the experimenter's regress, you have to do an experiment to find out: you cannot tell who the competent experimenter is or if the phenomenon exists. How do you know that a phenomenon exists? Do an experiment to find out. This is the regress Collins discovered in the middle of his gravitational radiation work. It was an important idea, I think, because the regress was then closed by different processes. If a claim was theoretically implausible, then theory, or rhetoric, or lack of funding, or just lack of interest would be used against it: it could be a variety of things.

I think Collins eventually formulated this in what he called the "Empirical Programme of Relativism" (EPOR), which had three stages. The first:

53. "Experimenter's Regress ... draws attention to the moments when, and the ways in which, skepticism is transformed from an analyst's concern to a scientist-actor's concern. The experimenter's regress is philosophical sociology rather than philosophy" (Collins 2002, 155).
54. Collins 1975.

another term for this controversy on the research frontiers is *interpretive flexibility*. That is an important term, because it crops up later in the sociology of technology. I don't actually know where it originated from, but certainly it was in the air. Collins and I talked about it, whether it was his term or mine: it was certainly a term that was in the air at the time. It simply meant that there was a clash over experiments or two different interpretations. So in his Empirical Programme of Relativism, a program set out in 1981,[55] he set out the first stage, which is to show the interpretive flexibility of experimental results. The second was to show how this flexibility was closed down, so you had this idea of closure or stability emerging: *consensus* emerging. This was called the second stage of the program, and that is where the term *closure mechanisms* was adopted, I think. The third stage of his program was to see how a laboratory fact was influenced or caused by the wider culture, the wider social culture or social forces. It would try and show how a fact on a laboratory bench is somehow being shaped by the wider world. The full goal would be to pass all three stages.

Most of the early studies focused mainly on the first two stages, and he edited this special issue of *Social Studies of Science* called *Knowledge and Controversy*,[56] where many of the early so-called controversies studies were all edited together: I've done mine on solar neutrinos,[57] David Travis on memory transfer,[58] Andy Pickering,[59] and Bill Harvey.[60] He released four studies, and Collins's own studies on gravitational radiation.[61] All of this was expressed four years later in Collins's book, *Changing Order*;[62] that is probably where all his case studies are written up—because at that time he had to write a book quickly, so this book came out. By then it was much

---

55. Collins 1981c, 1981d.
56. Collins 1981a.
57. Pinch 1981a. As Pinch points out, "scientific activity incorporates many craft practices and areas of tacit knowledge. It is the ill-defined and messy procedures inherent in craft activity which seem to provide the bulk of the ammunition for the criticisms documented" (145) in case of scientific controversies: "The lack of agreement over scientific certainty … is consistent with the view that the origins of assessments of certainty lie in the social world" (155).
58. Travis 1981.
59. Pickering 1981.
60. Harvey 1981.
61. Collins 1981b.
62. Collins 1985.

later, and Latour and Woolgar's book was the first book and got much of the credit in the field, although the ideas were kicking around among several people at the time. And then I came and got to work with Collins.

My history was that I had written a lot about Bohm, hidden variables, and studied that controversy. Turns out I was very lucky because Collins had this project on Uri Geller and the paranormal: he wanted to hire a research assistant, which was an amazing break for my career. In Britain these research assistants are usually like postdoctoral positions, they are faculty positions, but all I had was a master's degree from Manchester University; but he needed somebody who was working on controversies and who could interview physicists, particularly quantum physicists. Because you needed quantum physics to try to explain the metal bending, keys and spoons, by Uri Geller. I was the only guy in the world who could possibly do this! Unbelievable! He was interviewing all these postdocs with degrees and books, and suddenly this naive guy, Trevor Pinch, steps in saying, "Well, I've interviewed David Bohm. I am working on this wild idea of scientific controversies from the sociology of science perspective. I don't know what it all means, but this is what I am interested in," and I was just perfect. Collins said, "You have been trained with Whitley, who has got this particular approach, he doesn't really understand how to do this yet." Collins was just starting to understand because he had just made this discovery about the experimenter's regress. He said: "I'm going to train you up. We are going to write a book together[63] and I'm going to train you up on how to do this." That included how to interview scientists properly, how to set up a field work trip, how to do lots and lots of interviews. We interviewed together, he trained me up in interview technique: we were sitting in a room like this, we'd have a tape recorder like that, the old Sony cassette recorder; well, it's still a Sony … And he would train me. First of all, there were all the details of research: for example, how to buy the car that we were going to use to ride to the interviews, and sell it again to make money. How to fund the research, how to actually interview in a more aggressive way than I was used to, which he was very good at doing. At the same time, he taught me the key ideas that we needed in this field; I was really having a proper training with this guy.

63. Collins and Pinch 1982.

I have this funny story, which in hindsight is really strange, but just shows you how he sort of took me under his wing. One of the first things he insisted on was I had to dress differently. I used to be a kind of an old hippie, and I remember I was wearing women's clothes. I had my mother's coat. And he said, "If you want to work with me in California interviewing scientists, you have to get rid of that coat." So even my look mattered. I remember the first time we flew to California together: he leaned across to see what I was reading. I had a degree in physics, but I was mad keen on science fiction, so I was reading science fiction books. My book on the plane was Michael Moorcock, a science fantasy writer. I used to actually read his science fantasy and listen to the band Hawkwind.[64] I was really into it. "This is absolute crap, Trev. You can't. If you're going to work with me, you have got to start reading some decent stuff." So Collins gives me a whole list of literature, you know? It's good stuff, like Flann O'Brian, *The Third Policeman*, or William Faulkner. Because I had been a physicist, I actually had read some of these—or Hemingway—but not this literature in particular. He insisted that I had to read better and also said he was going to teach me how to write, because at that point in my career I was writing like a physicist. I remember he said we were going to write a book together. These were the days of typewriters, and he said, "Right, we're going to start doing this." He typed a paragraph: "Read this, Trev: what do you think? So, ready? Now *you* type something." "What do you mean? The next paragraph?" "Yes, just sit at my typewriter and type it now." So I typed it, and he went, "This is real crap." "Why?" "Look, this sentence is no good. Do it this way." He was actually teaching me the skill of writing, of which I am entirely grateful because a lot of learning is about having competence, and this guy was absolutely competent. He could train me to do all this stuff. The fact that I never got a Cambridge degree in sociology or any of these high-falutin' qualifications didn't seem to bother him. He actually may have preferred to have had someone a bit green and new. And I was interested in the same sort of issues, the sociology of science, to push this radical approach.

ST How fascinating: it's a method we have almost completely lost in our universities, at least in Italy ... It seems harder and harder to establish such

---

64. The science fiction and fantasy novelist Michael Moorcock (1939), mainly known for his *Elric of Melniboné* fantasy saga, collaborated on different occasions with the British psychedelic rock band Hawkwind.

a deep relationship that reminds me of the classic *discipulum-magister* relationship. But more in general, when I think about your career and your history, what strikes me most is that there was clearly a sort of strong bond between many of you working on these topics. Perhaps I would not call it a "school" or anything, but this network of people definitely appears to be not just an intellectual network but a network of meaningful relationships, and very strong ones, at that.

TP   Absolutely. Not only between us, our group. In Collins's group, there were three of us. It was very small: later I'll tell you a funny story about a conference—which brought Americans over for the first time—because we were so worried that we were so small, that we had to hold the conference in a different place from Bath University, where we were located, because we were worried that the Americans would say: "If this is so small, it can't …" It's curious, but it was literally Collins, a guy called David Travis,[65] who was his grad student, and me as kind of a research assistant. It was just the three of us doing this. It is unbelievable it was this small, but so it was. Then you had Barry Barnes, Donald MacKenzie, Steve Shapin, and David Bloor, up in Edinburgh, all headed by David Edge, who was kind of a father figure in the field. Edge was the editor of the journal *Social Studies of Science*. He played a really important institutional role in mentoring, getting our publications out, and helping us. He was an old Cambridge physicist, and he really helped: somehow he gave us legitimacy by editing that journal. And then Michael Mulkay in York University; a guy called Nigel Gilbert,[66] who worked with Stephen Woolgar,[67] who worked with Bruno Latour … Then there were all these people like Karin Knorr-Cetina. It was a very small field, and we often met up, invited each other to conferences and things. Personal relationships were crucial. Also, if you are lucky enough to be in a new field that is developing, with all the philosophers saying, "This is impossible. You guys are mistaken. You're radical relativists, you're wrong," you develop a strong feeling, like we did: "Hey, we're on something really important, a whole new view of science." But no one was taking us seriously back then. This was years before Latour was taken seriously, he still

65. See Travis 1981 on the scientific controversy over memory transfer by chemical means and on the replicability of experiments with planarian worms.
66. Among Gilbert's early publications more closely related to the sociology of science, see in particular Gilbert 1976a, 1976b, 1977; Gilbert and Mulkay 1980, 1984.
67. Cf. Gilbert and Woolgar 1974; Mulkay, Gilbert, and Woolgar 1975.

was this wild, weird French guy; Karin Knorr-Cetina was this incomprehensible German; Collins and I were often thought of just as undergrads who had physics envy—or someone who went running after physicists in this weird way—and Pickering was this old hippie who moved from physics to the field. Barnes and Bloor probably were the two who were taken the most seriously at the time because they were more established in a theoretical way. But philosophers hated them; they were constantly beating up on their works, saying they were impossible and wrong.

So there was a very strong sense among the people in the field, and it was helped by the fact that it was done in the context of Britain: we were in a small field anyway, nearly everything was done on a shoestring there. The most important thing is the idea that you are developing, rather than building big programs. In America, you build a big program, but there it was just, "Let's develop these ideas." We had the Strong Programme in Edinburgh, but the Edinburgh school was tiny as well, just a couple more people than the Bath School, not really any bigger. The whole thing was very small, and we had lots of strong personal relations, everyone knew each other. This is before the Internet, so we had to meet up face to face, at conferences and meetings, or PhD exams, or be on the phone constantly to each other. Our good pal, Bruno Latour, would often come across, give talks, sleep on my couch. I remember: I'd go up to Edinburgh, Steve Shapin[68] would take me under his wing, I'd go and stay at his place. I think it was like that for everybody in the field in the early days. There was this sort of a little wind of opportunity, when everyone really is doing the same thing, and it feels like: "Wow, we are onto something exciting here!" That was definitely the feeling with the sociology of scientific knowledge in the early days before everybody started fighting with each other ...

ST Yeah, SSK seems to be a very adversarial field, at least compared to mine, media studies, where usually people prefer to ignore each other than to argue. What was at stake in those early years?

TP At that time the main argument was between people in the lab and those working on controversies, and it basically boiled down to this: where is the best site for showing the social construction of scientific knowledge? That was the goal of our programs. The argument for us was that

---

68. The author, with Simon Schaffer, of *Leviathan and the Air-Pump: Hobbes, Boyle, and the Experimental Life* (Princeton: Princeton University Press, 1985).

controversies were the best site: the process of social construction of scientific knowledge does not happen in just one lab, it goes on across a wider number of laboratories. If you want to pick that up, it is better as a method to do interviews because you can be out in the field and follow the crucial actors, which are never all in the same site. So that's actually a weakness of *Laboratory Life*:[69] they are actually looking at this controversy over this molecule, TRF, but they only focus on one small set of actors in the lab, and some of the action takes place in other labs. They do not acknowledge this. Latour, and Lynch as well, have always said: "Oh, you pick up what's happening elsewhere from the microstudy in a lab." I don't agree. I think it is better to focus your study across a field because that is where the key processes are. It does not happen just in one lab. So that was the main argument with people like Knorr-Cetina, Latour, and Woolgar, who were focused in one lab while we were focusing on this wider field of scientific controversies.

There were also some differences between us and the Edinburgh Strong Programme which was based on historical case studies. MacKenzie's work[70] and Shapin's work[71] were similar in a way to what we were doing at Bath, but they were doing it historically while we were doing it contemporaneously. Shapin looked at phrenology and later with the historian Simon Schaffer at the controversy over Boyle's air pump's results,[72] which was a historical controversy. So, the difference was just about whether the research was contemporary or historical.

Then, the only other approach, at the time, was the textual approach, which was this discourse analysis of a very narrow sort of empiricist form that came out of York. First there was an article by Michael Mulkay and two students, "Why an Analysis of Scientific Discourse Is Needed."[73] Mulkay attacked an article by Collins and me[74] because we were not doing

69. Latour and Woolgar 1979.
70. See MacKenzie 1981a; MacKenzie and Barnes 1979. See also MacKenzie 1981b.
71. See Barnes and Shapin 1979; Shapin 1975, 1979, 1981.
72. Shapin and Schaffer 1985.
73. Mulkay, Potter, and Yearley 1983. See also Mulkay and Gilbert 1982, 1983; Mulkay 1981.
74. Collins and Pinch 1979. In particular, Mulkay, Potter, and Yearley criticize the distinction between "parapsychologists" and "orthodox scientists" methodologically assumed by the authors.

discourse analysis. Then there was a whole debate between Shapin,[75] Mulkay, and Gilbert about whether discourse analysis was an especially privileged form of data. Well, I think Shapin won the debate pretty conclusively, in my view, because it seems silly to claim (as Mulkay initially did) that somehow there was some kind of privilege about discourse analysis. Textual analysis is fine if there is just one thing going on because you pay close attention to the text, but to claim that it somehow has priority over all the other forms of data deriving from discourse is wrong. He did make that claim, and that was an extraordinary claim, but Shapin pulled the claim apart. Later on Mulkay was much more sophisticated—getting into reflexive analysis so that no form of data could ever be privileged. I have always seen discourse analysis as something like conversation analysis: it is fun, interesting, you can learn things, but you should never think that texts, or texts alone, or textual analysis, is everything. There is other stuff going on. I mean, one could have chosen a different method, and then someone has to get into what part of either method is the best. Anyhow, I think the controversy with lab people was more of a friendly disagreement. Mulkay invested more in it, claiming that was the only thing you could do.

So these were the main differences in the field. I remember going across the States early on and giving talks, saying, "You know, there are these different approaches: there's discourse analysis, there's controversy studies,

---

75. See Shapin 1984. Shapin points out how Gilbert and Mulkay oscillate between the proposal of an "inclusive" program of discourse analysis (inviting historians of science and other analysts to "describe and explain the situated variability of scientists' talk in addition to their usual projects") and an "exclusive" program, where "the goal of describing and explaining action and belief is said to be chimerical" (126). While welcoming their inclusive proposal, Shapin dismisses the exclusive one: "[Mulkay and Gilbert claim that] we cannot know action and belief through talk; all we can do is to 'reflect upon' the 'patterned character' of talk itself. They say that the discourse analyst 'is no longer required to go beyond the data.' This is the crucial point: discourse analysis is offered as a form of theory- and interpretation-free historical and sociological practice" (129) and, as such, it is assigned "'methodological' or 'analytical' priority over traditional exercises" (128). "I do not argue that Gilbert and Mulkay's interpretations are wrong, only that they are interpretations and not pure data. Discourse analysis is not, in this respect, qualitatively different from the sorts of practice Gilbert and Mulkay's restrictive program criticizes. Their recommended way out of our alleged 'analytical impasse' cannot be travelled" (129).

and there's lab studies." These were the main three forms. I remember that, in the early 1980s, it all seemed pretty clear. That is what the field was at that time. The ethnomethodology of Michael Lynch came later, but back then it was seen as more of a separate approach.

ST Since you are mentioning ethnomethodology, let's go back to the experimenter's regress, since this concept seems to resonate deeply with Garfinkel's ethnomethodology.

TP Actually, I think that was just a tangential interest to most people. The only person who was very influenced by Garfinkel at that point was Lynch, Garfinkel's own student.[76] He came into this field a little later, working away at lab study, but he was unknown at that time. We now know his dissertation was one of the earliest lab studies, but it remained unpublished for years and years. It even became a joke because it was always "Lynch forthcoming." We met him in Oxford, probably around 1981, and we had already been in the field for four years: a field like this is fast, four years in a new field is like a lifetime. In an old field, four years is nothing, but in a new field, four years is a long, long time. So Lynch came in later as a Garfinkel student, and Stephen Woolgar was the one doing something like ethnomethodology. Of course everybody had read about breaching experiments and had read *Studies in Ethnomethodology*,[77] but it was not very influential for us. If you look at Barnes and Bloor's early works, they were much more influenced by Wittgenstein, and Collins by Winch: you would not find very much about Garfinkel there at all.

ST I am rather surprised because, to me, the idea of the experimenter's regress and the focus on practices actually seems very close to an ethnomethodological approach. But it seems it came from other directions!

TP Well, I mean, Collins knew a little bit about it. Latour organized a meeting. We were doing our field work in California, my first-ever presentation in America, in 1976. I was at the Salk Institute with Bruno, and there was a meeting, and one of the people at that meeting was Aaron Cicourel,

---

76. Among Lynch's earliest works, see Lynch 1982, 1984, 1985a, 1985b. With Harold Garfinkel, see Garfinkel, Lynch, and Livingston 1981, Lynch, Livingston, and Garfinkel 1983. On ethnomethodology and social studies of science, see Lynch 1993.
77. Garfinkel 1967.

who had that book, *Method and Measurement*,[78] out. That had been influential, but it was more like a critique of standard sociology, something close to the interpretive tradition. Wittgenstein and Winch, with their inputs, were more influential on our group and on Bloor than Garfinkel's *Studies in Ethnomethodology*. So it was interesting, but it came about later.

ST   I was thinking about ethnomethodology because a scientific controversy seems to me a sort of "naturally produced" breaching experiment.

TP   Here is an interesting thing: exactly that phrase, "naturally produced breaching experiment." That is exactly how Collins later presented it![79] He actually talked about it like that, and I use it to this day in teaching—I use this as an example in teaching controversies. But that came later. We were already studying controversies when we started to think, "Why is it that controversies are so good?" David Edge had written something about controversies: getting at the underbelly of science[80] so as to make the implicit explicit. The underbelly of science, where everything comes out—but then Collins actually came up with the idea. Collins, by the way, had met Garfinkel at this point and came out, then, presenting controversies exactly as naturally occurring breaching experiments; he said that when a controversy comes along, rather than you, as a sociologist, having to do the breaching, the scientists do the breaching for you, and you just sit back and watch what happens. So, Garfinkel. But it's interesting how the Garfinkel way became a way of expressing and legitimizing it but not a method for doing it, because we were already doing it. We were more like, "What a beautiful way of expressing it." I do it to this day. We often repeat this physics metaphor: "Physicists learn about systems by punching systems." So if you have a billiard ball on a table, a pool ball, look at it: you know nothing. You bounce another ball off it, and you immediately learn about laws of conservation of momentum and energy, and we were

78. See Cicourel 1964, where Cicourel addresses the limits of mathematization and statistical analysis in sociology. Initially associated with Garfinkel's ethnomethodological approach, Cicourel will refuse the identification of interaction and verbal accounts, branding his declination of ethnomethodology "cognitive sociology." See Cicourel 1974.

79. "Disagreements among actors serve the same function for sociologist-observers as the deliberately engineered breaching experiments described by Garfinkel" (Collins 2003, 671).

80. Edge 1976, quoted in Latour and Woolgar 1979.

the same. We had to find a way of breaching science, and the answer was, of course, "naturally occurring breaching through controversy," but this is only a rationalization for explaining: a teaching tool rather than a method (which only came later).

Garfinkel appeared on the scene later. I remember meeting him, he was with Lynch and John O'Neill,[81] and it was kind of interesting. He hadn't read anything of what we were doing; he was framing it in terms of Gerald Holton[82] reacting against Merton. So Garfinkel was framing it all like, "Well, we're going to do ethnomethodological studies of scientific work, to answer somehow what Merton is doing." At that point he seemed completely unaware of the fact that a whole sociology of scientific knowledge, from a very different tradition, had just been launched.

ST   So you already had the full-fledged notion of the experimenter's regress before framing it in Garfinkel's terms. What are its implications for the production of scientific knowledge? How does it deconstruct the standard view of the relation between theory and experiment?

TP   Actually, the experimenter's regress is only one of a number of ideas to understand the deconstruction of the relation between theory and experiment. There was an idea, coming from the philosophy of science, that helped better: *theory-laden observation*. It is an idea that goes back to an Oxford philosopher called Norwood Russell Hanson,[83] who asked this sort

81. John O'Neill is distinguished research professor of sociology at York University, Toronto. In his early work, O'Neill contributed to the development of ethnomethodology through a systematic dialogue with the phenomenological tradition.
82. Born in 1922 in Berlin and having immigrated to the United States in 1938, Gerald Holton is a physicist and historian of science who is now *emeritus* at the history of science department in Harvard. In 1972 he founded *Science, Technology, & Human Values*, which later became the official journal of the Society for Social Studies of Science. Holton refuses positivistic accounts of science, showing how scientific imagination is shaped by a recurring and restricted set of "themata" (and "anti-themata"), preconceptions "that are not resolvable into or derivable from observation and analytic ratiocination" and that therefore can originate controversies over the interpretation of the same observational data (Holton 1973, 1978, 1986). Pinch published reviews of two of Holton's books: "The Advancement of Science, and Its Burdens, by Gerald Holton," *Times Higher Education Supplement*, May 15, 1987; and "Thematic Origins of Scientific Thought: Kepler to Einstein, by G. Holton," *Times Higher Education Supplement*, Aug. 26, 1988.
83. See in particular Hanson 1958, 1969.

of question: "If you take, say, a crater on the moon. If you have the idea of a crater on the moon, doesn't 'crater' itself involve some theoretical idea? It's going to be round and pitted, and you have some ideas about how that thing was formed."[84] So the theories are embedded in observation: *theory-ladenness of observation*, a very well-known concept in the philosophy of science.

There is another closely related philosophical way of expressing it: it is the Duhem–Quine thesis in the philosophy of science. It says that it is impossible to test a hypothesis in isolation because it would, anyhow, imply other auxiliary hypotheses. Every observational statement in science has embedded in it other clauses that you would call *ceteris paribus* clauses; everything else is background assumption. Think about a gold-leaf electroscope, measuring static electricity. It looks like a kitchen scale, there is a plate there; you bring up a balloon, and the two gold leaves will open to show that static electricity is passing through the plate to the gold leaves, and the charge is pushing them apart. But we must assume as a background that there is no dampness in the air because, if you get no opening of the leaves, it might simply mean the plate has lost all its charge via water moisture. So the measurement of the charge depends on the background assumptions. One of the background clauses, theories if you like, to measure static electrical charge correctly is that there is no dampness in the air. So all observation statements involve these background theories. This point was established by Imre Lakatos, by Quine,[85] by Duhem.[86] It's called the Duhem–Quine thesis. So those are the philosophical ways of dealing with it, and among the sociological ways, the experimenter's regress is one.[87]

84. "An example of how an observational term is theory-laden is provided by Hanson in his discussion of Galileo's observations of 'craters' on the Moon's surface. Hanson writes: 'Galileo often studied the Moon. It is pitted with holes and discontinuities: but to say of these they are craters—to say that the lunar surface is craterous—is to infuse theoretical astronomy into one's observations. ... To speak of a concavity as a crater is to commit oneself to its origin, to say that its creation was quick, violent, explosive' [Hanson 1958, 56]." "The term 'crater,' then, is not purely an observational term but has embedded in it theoretical notions" (Pinch 1985c, 12).

85. Quine 1961.

86. Duhem 1954.

87. Pinch 1985c, 14:

During observational disputes challenges to the background assumptions which have gone into the observation are made. Because the results are only as good as the assumptions used to get

I actually have found in my own teaching that a quicker way to get at it is through the example from my own PhD on the solar-neutrino experiment, where I wrote a paper on the externality of observation,[88] which is one of my better received papers actually. It has got a large number of citations. People have told me that. Well actually, Collins, he would tell you if you interviewed him, "Pinch had two OK papers in 1984." One was that one on the social construction of technology[89] (which wasn't so important, he would say!), but the one on the evidential context and externality of observations was what he would call a real discovery, a really important paper. So I often use that as an example of the relationship between theory and experiment. Because theory-ladenness, as expressed by Norwood Russell Hanson, doesn't quite get it because he has this pictorial visual idea of things, of a crater. Most scientists in modern science are not looking at pictures like that, they're looking at data that are actually like the stuff Latour talked about: inscriptions. The outcome of the solar-neutrino experiment is

---

them, to doubt the assumptions is, in effect, to doubt the results. And … there are plenty of candidate assumptions which can be called into question. Philosophers of science are more than familiar with this property of scientific observation through the Duhem–Quine thesis. Since observational statements depend on a network of assumptions, the logical force of a recalcitrant observational statement can always be removed by a change elsewhere in the network. Essentially, the sociological studies describe scientists' attempts to remove the force of recalcitrant observations by challenging background assumptions in the observational process.

88. Pinch 1985c:

It seems that in scientific observations … our internal biological receptors have become "externalized." That is to say, the process of observation in modern science is one in which experimental practices and theoretical interpretations take on central importance. … By this process of externalization, observation becomes a question of studying a chain of surrogate phenomena via a series of manipulations and interpretations, and this highlights the fundamental ambiguity over just what has been observed. (8)

The term "theory-laden" is, however, not directly equivalent to the term "externality." … Externality is defined in terms of the amount of the observational situation which must be encompassed in compiling the [research] report. The more external the report, the more assumptions about the observational situation that must be included. … Many such assumptions will be embodied in the craft practices for operating this particular piece of experimental equipment. … We can say that reports of high externality are heavily theory-laden, but we are then using "theory" in a very special way. As has been emphasized, high externality reports often incorporate assumptions about the working of the apparatus at lower degrees of externality. (13)

89. Pinch and Bijker 1984. The SCOT approach, presented for the first time in this paper, will be addressed in chapter 3.

actually an inscription device.[90] It is a form of sophisticated Geiger counter, called a multi-proportional counter, which produces a graphical output of energy versus speed of particles: it has the energy of particles on one axis and the speed on the other, and you have this box where the key particles (evidence of neutrinos) are. They are just splodges. So you can use this "externality of observation" notion I developed in that paper, and "evidential context,"[91] to think: "How do you go from the Sun to these splodges? What do scientists actually look at? What are the scientists observing?" They're just observing these ... splodges. And how do you make sense of these splodges, in terms of what's going on in the middle of the Sun, in terms of neutrinos? And the way that that works is by a chain of surrogate phenomena where splodges stand in for argon (Ar^{37}) decays, which stand in for chlorine (Cl^{37}) atoms, which stand in for neutrinos, which stand in for solar neutrinos: you can call it a "translation," or you can say it is a process where more and more of the observational situation and has been incorporated, and more and more externality.

Basically, you also have to use assumptions, including practical assumptions, as well as what you might want to call "theoretical assumptions,"

90. Latour and Woolgar 1979, 51:

An inscription device is any item of apparatus or particular configuration of such items which can transform a material substance into a figure or diagram which is directly usable by one of the members of the office space. ... The particular arrangement of apparatus can have a vital significance for the production of a useful inscription. ... An important consequence of this notion of inscription device is that inscriptions are regarded as having a direct relationship to "the original substance." The final diagram or curve thus provides the focus of discussion about properties of the substance. The intervening material activity and all aspects of what is often a prolonged and costly process are bracketed off in discussions about what the figure means. The process of writing articles about the substance thus takes the end diagram as a starting point. Within the office space, participants produce articles by comparing and contrasting such diagrams with other similar diagrams and with other articles in the published literature. (Latour and Woolgar 1979, 51)

91. Pinch 1985c, 10:

When talking about observation in science one must always be aware that observations are usually made for some purpose, such as, for example, to test a theory or confirm another observation. One way of describing this is to refer to the evidential context of observations. The evidential context can include such things as a body of knowledge, a theory, law, or hypothesis, a classificatory scheme or one class of objects, a set of observations or one observation. The relationship between the observational report and the evidential context defines the evidential significance of the observation. ... Observational reports can be said to be "relevant" only in terms of the evidential significance that can be ascribed to them. However, there is one very interesting property associated with evidential significance. That is that observational reports may take on significance in a variety of different evidential contexts.

to move in the translation chain from those splodges on the graph to what is going on in the Sun, at different levels. So you have to go from these splodges to the Ar^{37} decays: this involves statistical theory, and it also involves assumptions about how well the apparatus worked. Some atoms of Ar^{37} have been extracted from a huge detection tank: so, was the procedure of the experiment followed correctly? You need to make assumptions about the experimental procedures, assumptions about how neutrinos interact with Cl^{37} atoms producing Ar^{37}, which also involves theories of physics. So what you get here is a much richer picture of what theory-ladenness means. This connection between theory and experiment is something I have written a lot about, and this sort of example teases it out better. So, rather than talking about theory and experiment, I talked about how evidential significance is attached to particular observational statements: what level this observational statement could be pitched, since it could be pitched at different sorts of levels—I call this "externality"—and how these levels change during a scientific controversy.

It is a very, very simple point. Basically, if you are in a scientific controversy, if you make more general statements about your observations (which means going closer, more proximal to this view of them being splodges), you basically say things less theoretically, but people are more likely to accept your results. So in this controversy about solar oblateness,[92] when Dicke said, "Well, I am observing excess brightness," no one could doubt that. He is basically making this move: he is not saying his experiments were wrong. He is making this move, going a bit nearer on this chain of externality. The way these things move backwards and forwards on this chain is the thing that I discovered: these different levels and how to think about that, in terms of how the different parts of theory at stake are embedded in the apparatus. That is how I think about theory and observation.

So it's not directly linked to the experimenter's regress, except that when you have a controversy and you have a regress, then all that situation moves. But the experimenter's regress is not a fine-grained enough idea, yet. It does not tell you anything in detail about the relationship between theory and experiment because the regress may be closed by all sorts of

92. Robert Dicke's attempt to measure the oblateness of the Sun in 1964 is the second, comparative case study in Pinch 1985c.

things: the theory may be part of it, but it may be the experimental procedure. So it is not a fine enough tool to answer the question you were asking about the relationship between theory and experiment. I think what's so powerful about the experiment's regress is that it gets at this idea of skill, which is underlying a lot of the work that Collins and I have done and which I think is still poorly understood and understudied because very few people investigate what counts as a skill.[93] And this is important in sound studies as well: to think about skills, the skill of listening, and also the skill of practicing. You know, the terms *habitus*, the term *skilled*, whatever that is … these "communities of practice" of Lave and Wenger.[94] Garfinkel and Goffman may also talk about it in some sense: these are the things I am interested in at the moment.

1.4 The Bath School: Methodological Relativism

ST As we will see, the conceptual framework you turned up working on science will be very influential for your subsequent works on the social construction of technology: you will conceive technology as the object of a controversy. Yet, there is another central point we have not discussed so far, and it is the peculiar declination of relativism that the Bath School was adopting at that point: what you call "methodological relativism."

TP Yes, relativism, and it is also important to ask what sort of relativism Collins and I meant at this stage because there's a lot of confusion about this. Early on, Collins was a much more radical relativist. He later became what he stipulated was a "methodological relativist." Methodological relativism meant that in studying controversies where you do not know the outcome, it made sense to act as if you did not know what the outcome was. It was a very simple point, the point of anthropologists going into the field: don't prejudge the phenomena, just wait and see. And we didn't have any strong prejudice one way or the other: ours was a kind of distancing approach. That was what he meant by a methodological form of relativism.

ST It is like trying to assume the points of view, the discursive positions, of the different actors involved in the controversy, before one of them, in

93. On skills, see also Pinch, Collins, and Carbone 1996.
94. See Lave and Wenger 1991.

hindsight, becomes the one who was right: which is not easy if you are dealing with a historical or an already-settled controversy.

TP Exactly. Before that, with *Frames of Meaning*, I think he had been a stronger sort of relativist. I think I can remember him saying things like, "There is a group of people who constitute the phenomena of the paranormal, and there is a group of skeptics, and both groups exist, and we cannot say any more on that." I think he was what we might call "a fully fledged epistemological relativist" back then. But I remember him changing his mind, to a certain extent, because he thought that it basically made no difference, because you could never prove relativism as a position anyway. So it didn't make any difference if you couldn't prove it. Many people hated epistemological relativism: it would be an easier position to defend, just to be a methodological relativist. I think that is why he gave in, in conclusion, since you could not prove it anyway. What was the point of having a big fight over something that could not be proven?

ST It makes sense, even though one of the criticisms that would be leveled against you and SSK some years later, especially by scientists, is that it is impossible to be a methodological relativist without being an epistemological relativist.[95]

TP Yes, that was another issue, but I think you can be a methodological relativist, and it does not necessarily follow that you are a stronger sort of relativist, like an ontological relativist. Because a methodological relativist can say something like what we say in *The Golem*[96] book. There is actually a theory of truth in *The Golem*: "the consensus theory of truth." Eventually a consensus will emerge in the scientific community as to whether these phenomena exist or not, and that consensus means you can think of some phenomena as the current scientific truth of the matter. Then it makes sense to act on that, as the consensus has been achieved in the scientific community. You do not need to buck that consensus if you were doing a special sort of project, but for all intents and purposes, that consensus *is* the way the world is now. One can still study controversies with a methodological

95. This position is expressed, for example, by Jean Bricmont and Alan Sokal (2001) and several other scientists contributing to Labinger and Collins 2001, the volume that represents one of the key episodes in the so-called science wars between the SSK community and scientists (see sec. 2.2).
96. Collins and Pinch 1993.

relativism. It does not mean to deny that consensus has been reached in the long run. The consensus is, now, that high fluxes of gravitational radiations do not exist. The fact is that there are no firm epistemological grounds for knowing that. It is a mixture of things, but the consensus has emerged in the community.

ST Yet, your critics have argued that this line of argumentation somehow "circumvents" the problem of relativism since what really makes a difference is how "nature" counts in that mixture of things that constitute the process of building consensus. The fact is that if "nature" does not count at all, then it seems actually quite hard to tell methodological relativism from more radical kinds of relativism, because they would have the same consequences for our explanations—as your critics have pointed out. On the other hand, if we admit that nature plays a role of any sort in that mix, then we still have to clarify if, and how, that role can be accounted within a sociological perspective, or at least how can it be integrated within a sociological account that maintains the principle of symmetry.

I am trying to clarify this point because I think that these are the roots of another controversy that we are going to address: the one that, at a later point, will divide the STS community over the issue of nonhuman agency. As we will see, the actor-network theory will try to answer to this challenge by acknowledging agency to nonhumans, even if only and always within a network of relationships between heterogeneous elements. Similarly, Pickering[97] acknowledges to nature a sort of resistance when dealt with through scientific practices. On the other hand, the Bath School stresses the relevance of the "interpretations" of social actors: nature's agency is "always interpreted" and included in a representation, notwithstanding the centrality of practices and tacit knowledge for the approach.

This stress on interpretation brings us to semiotics and, in particular, to the debate over deconstructionism[98] that seems to have several points in common with the SSK debate over relativism and nature's agency. That debate reached its peak in the late 1980s and early 1990s: at that time, semiologists had already clarified how a large part of the "meaning" of a text is not determined by the text itself (say, by its "structure") but is coproduced by the interaction between specific textual devices and the

97. See in particular Pickering 1995.
98. For an overview of the debate, see Eco 1992.

readers' *encyclopedia* of knowledge and practices of decoding. On the other side we had the deconstructionists, like Jonathan Culler[99] or Stanley Fish:[100] brutally simplifying, for them meaning was completely contextually constructed. Some years later, in his book *Kant and the Platypus*,[101] Umberto Eco has summarized his epistemological position: he admits that nature and texts are underdetermined, and, as such, they are prone to interpretative flexibility. Yet, the fact that nature—or texts—can be interpreted in many different ways does not imply that they can be interpreted in whatever way. Some interpretations are blocked, or better, as he writes, "resisted." Nature and texts do not impose readings but "resist" some of them,[102] even if this capacity of nature to resist is just, of course, an effect of our interrogations. This position seems to avoid radical epistemological relativism while allowing some space for a sociology of (scientific) knowledge. Which are the affinities and the incompatibilities of this stance with the school of Bath?

TP Well, that would block radical relativism because you would be giving some agency to nature to do certain sorts of things, to "block"

99. See Culler 1982.
100. See Fish 1980.
101. Eco 2000 (orig. ed. *Kant e l'ornitorinco* [Milan: Bompiani, 1997]).
102. Eco 2000, 52–54:

We use signs to express a content, and this content is carved out and organized in different forms by different cultures (and languages). What is it made from? From an amorphous stuff, amorphous before language has carried out its vivisection of it, which we will call the continuum of the content, all that may be experienced, said, and thought ... Hjelmslev called it ... "purport." ... What does it mean to say there is "purport" before any sensate articulation affected by human cognition? I would prefer to translate Hjelmslev's meaning as "sense," a term that can suggest both meaning (but there is no meaning or content before a given language has segmented and organized the continuum) and direction or tendency. As if to say that in the magma of the continuum there are lines of resistance and possibilities of flow, as in the grain of wood or marble, which make it easier to cut in one direction than in another. ... If the continuum has a grain, unexpected and mysterious as it may be, then we cannot say all we want to say. ... Here we should avoid a misunderstanding. When we talk of the experience of something that obliges us to recognize the grain and lines of resistance, and to formulate laws, by no means are we claiming that these laws adequately represent the lines of resistance. If along the path that leads through the wood, I find a boulder blocking the way, I must certainly turn right or left (or decide to turn back), but this gives me absolutely no assurance that the decision taken will help me know the wood better. The incident simply interrupts a project of mine and persuades me to think up another one. ... Language does not construct being ex novo: it questions it, in some way always finding something already given (even though being already given does not mean being already finished and completed). ... This already given is in fact what we have called the lines of resistance.

certain sorts of positions. It looks to me like a philosophical formulation of the issue, because the question would become, "In a particular scientific controversy that we'd be studying, what would nature be doing in terms of blocking?" If you thought the moon was made of green cheese, you could say that nature could block that, because it does not seem to be made of green cheese, from all the conceivable tests that we are doing on nature. But the trouble with that is that it tackles the issue in abstract terms. If you looked at that in terms of a scientific controversy, people who pro-pose radical views like "the moon is made of green cheese" would never be taken seriously because that is such an implausible claim, and no test would be actually made. So it's hard to say that nature alone is doing the blocking. What I mean is that any discursive context where you put these arguments immediately makes them messier. So those sort of examples, like Eco's, only work as a thought experiment, but when you have real historical actors—or sociological actors, in a contemporaneous case study—and you see the sorts of beliefs that could be "blocked" by nature, it is impossible to say a priori which of those particular beliefs about nature were blocked by nature in any particular moment. There can be doubt and disagreement over them. You can then say, "Nature's done the blocking" in cases after the fact, post hoc. Well, that is the problem, except when you come out with these philosophical, implausible cases. When you translate the rejection of those views into an actual historical position, the blocking would quickly be seen as a social process as well as something strange about nature, you know, as giving some agency to nature. My opinion is that it would be hard to disentangle.

ST Clearly there is a problem with disentanglement, the problem of accounting for that mix as a mix, and you clarified the problems that arise when we try to account for nature's agency in the mix that brings a contro-versy to a closure. Yet, pushing a bit further this confrontation with semi-otics, in Eco's terms we could say that since a controversy as a discursive context—as a frame—is defined by an uncertainty that is first of all shared, the involved actors are performing their conflicting interpretative efforts within a space of maneuvering where "resistance" appears, at least tempo-rarily, weak. In this sense, for a semiotic-inspired approach, the point you made would be acceptable as it gives nature a strange and limited kind of capacity to act, or to resist. The resistance of nature would be identifiable, first of all, in the conditions that led to a situation of shared uncertainty.

Here Eco returns to another of those "thought examples," as you have just defined them: he writes that if you find a boulder blocking your path, it would be controversial if you have to circumvent it from the left or from the right, but it would suffice to persuade you that a project of yours, say, one going straight, needs a revision. By the way, Harry Collins also uses a very similar example, though with opposite implications: for him the "rock in our path" is evoked as paradigmatic of what seems to instruct us without needing an interpretation.[103] In the second instance, nature's "resistance" would be identifiable within the field of shared uncertainty, in terms of constraints to the different moves available to social actors within the controversy.

To keep it less abstract, I would go back to one of the case studies we have already discussed: the controversy on gravitational waves.[104] Back then there was not a consensus on whether what were being experienced at such a high level were gravitational waves or not. This lack of consensus was triggered by a mismatch among theory and experimental results that, first of all, called for an interpretation. Furthermore, we know that the conflicting interpretations of the high level of gravitational waves registered were many, and none of them alone could close the controversy. But the fact that they were many doesn't mean that any of them was possible. For example, nobody could claim that what Weber was measuring came from the wind since his antenna was set up in a vacuum chamber. In this case, it seems that "resistance" was not enough to close the controversy, but enough to reduce the argumentative moves available to the players involved in the controversy.

TP That is like the green cheese argument. The reason that people do not claim that is because, given a discursive context, it would just seem incredible. It is an implausible argument; just a scientist suggesting something implausible. Now, if they suggested it in all seriousness, then it would

103. "As we stumble against a rock we do not seem to have to think about obeying its instructions. It will give us guidance about where to walk in its vicinity whether we think about it or not. We may walk beside it or away from it, but not through it. We do not have to decide not to walk through it. Our actions are caused directly by the rock rather than by our interpretations of what the rock is. ... A rock instructs everyone equally and obtains uniform results. Rocks seem to be cultural universals" (Collins 1990, 50).
104. See Collins 1981b.

become a very interesting argument. In other words, if one of the leading scientists had suggested that the results of these experiments were due to the wind, the controversy would have turned in that direction because there could be an explanation. You could say, as a plausible explanation, that the wind had rocked some experimental phenomenon that was misinterpreted as a vibration in this antenna, in the shaking bars that Joseph Weber had looked at. Trucks were actually proposed; highway trucks heavily rolling on the highway were causing the vibrations. So there are things that seem kind of crazy on the face of it but can be taken seriously according to the discursive context.

I think that's a profound point: in a real scientific controversy, one cannot say in advance which of these factors are going be the relevant ones. Some will be more plausible than others, and you can always imagine some implausible ones. But that is an imaginary game: no real scientist has actually advocated that position. So then, what is the status of those imaginary claims? They cannot play a part in the controversy. You are just saying, "That is what we imagine would be an implausible thing." And that doesn't seem to carry that much weight with me. Because we are not involved in this controversy, we are not experts about it. That's exactly why I think that every time one gets an issue to do with, on what part of nature is, it is a methodological rule to insert it back into an actual discursive context, where people are arguing about these things, and then see.

Here's the thing: one can look at these controversies, as I have looked at solar neutrinos in my book, *Confronting Nature*,[105] or Collins looks at gravity waves, and you can see, at a certain point, that the scientists in theory could have pushed their arguments more but did not. In other words: this scientist involved in the solar-neutrino controversy, Jacobs, claimed that Davis's results were explicable by experimental error. Davis had found a certain number of neutrinos: the "wrong number," in inverted commas, as it was not the number predicted by standard theory. Jacobs said that, basically, some of these neutrinos were being lost—trapped in the detection tank. It is a technical issue, but some neutrinos would have been trapped by the argon in the tank, so everyone was counting the wrong number. Davis did a test to show that this looked pretty unlikely: he used something that is very much like Ar^{37}, but was actually Ar^{36}, and he said, "Look, I can count

105. Pinch 1986. See also Pinch 1981a, 1981b, 1985b.

the right number." Now, at that point Jacobs decided not to challenge Davis's procedure any more. At this point you could say that nature has blocked this explanation. Okay? Now, in our work, Collins and I pushed on these sorts of things: "So, what has constrained Jacobs?" At this point, one could imagine that Jacobs could push the argument further by saying, "Actually there is a difference between Ar^{37} and this Ar^{36} that you've used in this test experiment." This difference explains why nature has not in fact blocked you. So you are not forced logically to say that nature has blocked. This is a very nice example, but the real reason for the block is that culturally, at this point, no one would accept Jacobs's view that these things were significantly different. That is something that's bound up in the culture, on what counts in things as being similar and different.

So even in these hard cases, you can never say for sure that it was nature that forced. You were forced by the consensus about what is plausible to challenge. And there is the same in Collins's work; at one point he talks about Weber having accepted a certain calibration of his apparatus.[106] It is a special, electrostatic calibration. Weber did not have to accept that. He would have been challenged a lot more if he had not accepted it, but he *could* have challenged it: he would have been culturally more out on a limb, but it wasn't nature that would have forced him out. It was culture, because he had just started to become too deviant. So the theory is that you can push this. It is very hard to say at which point nature is really acting. And that is the way we analyze these controversies. So when you get nature acting for sure, it always comes in like a post hoc construction. You know, Latour also made this point with his Janus-faced figures in his book, *Science in Action*. What looks like the actions of humans before the controversy,

---

106. Collins 1985, 105:

Calibration is the use of a surrogate signal to standardize an instrument. The use of calibration depends on the assumption of near identity of effect between the surrogate signal and the unknown signal that is to be measured (detected) with the instrument. Usually this assumption is too trivial to be noticed. In controversial cases, where calibration is used to determine relative sensitivities of competing instruments, the assumption may be brought into question. ... Weber, in accepting the scientific legitimacy of electrostatic calibration for his gravitational antennae, thus accepted constraints on his freedom to interpret results.

Collins has been documenting the ongoing research of gravitational waves from a sociological perspective since the early 1970s. His most recent volumes on the topic are *Gravity's Shadow* (2004) and *Gravity's Ghost and Big Dog* (2014), and *Gravity's Kiss* (forthcoming, MIT Press).

looks like the action of nature after the controversy has been settled. When agreements have been reached, if I came to push them, then it would look like nature.

So the argument by Umberto Eco that "it resists but doesn't act" sounds like he's wiggling out of the problem. Because to resist is to act, actually. Not acting is acting! This is a crucial point because, when you get into why a scientist like Jacobs does not push forward, one could imagine (supposing Jacobs, at this point in his career, had not lost tenure—and he did not get tenure, he was a very marginal figure) that he would have been in a stronger position, and pushed it further, if he had more credibility. Therefore you are in the realm of social resources, credibility, and culture. At that point, it was not clear whether nature was acting unambiguously. This is the methodological relativist position; while analyzing these controversies, one cannot say in advance what nature is doing. That is the thing. But at the end of it, once it is over, then you can say a consensus has been reached. When Jacobs lost this consensus, then you could have started to say "This is how reality looks." But in terms of methodological relativism, it is impossible to say that nature is acting at any particular point. That is how I'm trying to make clear what the position is.

ST   There's a last point that has been opposed to methodological relativism and that I'd like to address: we said that controversies are specific discursive contexts characterized by a shared uncertainty, but also that they can be conceived as "naturally produced breaching experiments" and, as such, are suitable to tell us something about science in general. Don't we risk conflating methodological and epistemological relativisms here? I mean: extending what counts in this specific discursive context, defined by a controversy, so fragile and unstable, to science in general?

TP   We answer this objection in the afterword of *The Golem*,[107] since people have criticized the book for focusing too much on controversies in science. That's our answer: "Of course we are not saying all of science is controversial, far from it; but we say that the sort of science that the public does not understand is, in fact, controversial science. This is the science that leads to instability, and when you have a controversy like the ones we talked about,

---

107. *The Golem* (Collins and Pinch 1993) is the first of a series of volumes by Collins and Pinch aiming to foster the public understanding of science and technology and to popularize the main acquisition of SSK. The *Golem* trilogy will be addressed in more detail in section 2.1.

this is the sort of science that people need to understand." This justification on why we focus more on controversy is valid in the world of public understanding of science. When I was doing it as a researcher (before I did *The Golem*), it was because it was a method that seemed to reveal more about what was going on in science, a very powerful method, because we have all these ideas, just like the experimenter's regress, for looking at controversies. If you never looked at controversies, in my view, you would never find certain interesting stuff. I still think this is true. At that time there was not a strong sense of framing it in terms of order and stability, searching for fragility or the breaching mechanisms. That was not the way it was framed back then. As I told you, Garfinkel came at a later point. We saw controversies more like good research sites. I really saw that in empiricist terms: "I am learning a lot from these controversies about science."

ST   To conclude this part of the conversation on relativism, I am still curious about one thing. Reading about the quarrels you had with philosophers back in the 1980s, I was really surprised that relativism was seen as somehow connected with extreme right political positions—even with Nazism, as you write in *Confronting Nature*.[108] This is quite surprising for me, since nowadays we tend to associate relativistic stances with leftist positions of the postmodernist kind.

TP   Yeah. Well, the historical roots of this come from the generation of refugees from Nazi Germany. I could just mention one name: Gerald Holton, who was a very important historian of science institutionally. He was the chair of the Harvard History of Science Department. I don't know if he actually was a refugee from Nazi Germany, but his generation's claim was that they had seen the effects of relativism, extreme relativism, in Nazi science. For them it was a real political issue. I also remember a philosopher of science here at Cornell, Richard Boyd, a realist philosopher of science, telling me that in the political struggles at Harvard University in the 1960s, the relativists were always on the right, and the realists were more left-wing. It was like being guilty by association: they thought that with relativism you cannot distinguish one program from another and all programs are equal. Someone who accuses Einstein of being Jewish might make the point that

---

108. "Relativism is still an unpopular doctrine because many people equate relativism with a mindless subjectivism or irrationalism, or even worse, with Nazism" (Pinch 1986, 10).

there is a Jewish physics, and he would seem to have got much legitimacy, you cannot condemn it. That is the sort of position they had. They were really worried about that.

Of course, extreme relativism and the rise of extreme right-wing politics are a very legitimate concern to worry about. We have seen that in America in a new form, with antievolution arguments coming from the right wing as well. So I think it is something to take seriously and worry about; but it was kind of strange for us as well because most of the people who were relativists in Britain, like David Bloor, Harry Collins and myself, Stephen Shapin, and Barry Barnes, were actually politically on the left. Malcolm Ashmore wrote quite a nice little article on the rhetorics of this, called *On Death and Furniture*.[109] Ashmore is an interesting guy, a sociologist of science who did one of the best reflexive studies on the field. He went around and interviewed the members early on and wrote this book, *The Reflexive Thesis*.[110] I was one of his PhD examiners at York. He later wrote *On Death and Furniture* at Loughborough University, with Derek Edwards and Jonathan Potter, about the rhetoric of these arguments about relativism: basically, it's about how sooner or later somebody will start banging the furniture, banging the table, and say, "Look, you can't deny that this is real." When you walk into a table or something ...

ST   ... like the boulder in Eco's example!

TP   Exactly. And death is another of the things you can't deny. Also, you cannot be a relativist at 40,000 feet or 50,000 feet because you wouldn't be on a plane, and if you are there it is because we all know that airplanes work, otherwise you would not step in it or you would be dead.

ST   There is a recent paper by Bloor against this reasoning.[111]

TP   Yes: death is a dramatic example. Ashmore does a nice job deconstructing those arguments, showing how they work rhetorically, and Nazism is one of them. I have written about that in a paper called "Relativism: Is It Worth the Candle?" in a special collection for Barry Barnes,[112] where Bloor has that paper on relativism at 30,000 feet, that has a different take; he is more of an epistemological relativist than I am. The book came out in 2008,

109. Edwards, Ashmore, and Potter 1995.
110. Ashmore 1989.
111. Bloor 2008.
112. Pinch 2008a.

but the paper had been presented and read long before that. I originally presented it at a conference where I had a debate with a realist philosopher of science called Phil Kitcher, in New Orleans.[113] I had a big argument with him in a session on relativism, and that is the last thing I've written on relativism. Basically, I am saying you do need methodological relativism, but I wouldn't go further than that. By that time I had also had many fascinating discussions over these sorts of issues with an anthropologist friend of mine, Richard Rottenburg,[114] whom I first met in Berlin. Richard, by the way, is one of the few anthropologists who really gets STS. He originally worked with Bernward Joerges, and Richard and I once taught a class together at the Viadrina University in Frankfurt under Oder in Germany, on the Polish border. He and I have a famous unpublished paper together on "reverse flows," which is about studying technology in a postcolonial context. Richard, who did his fieldwork in the Sudan, thinks about issues concerning relativism in a very consequentialist way—what if you or your family get sick in a remote part of the Sudan? Are you a relativist about indigenous medicine then? I think not! I haven't written again on relativism because it doesn't seem to be that big of an issue anymore, no one really bothers with relativism these days so much. No one is accusing me anymore of relativism when I give a talk on science.

ST Yet, I think that if you do not keep this debate in mind, it becomes hard to understand what is at stake in the quarrel over nonhuman agency …

TP In fact, when we're going to get onto the movement of these arguments over to technology and to the social construction of technology, it is important to stress that what Bijker and I[115] intended to do was not really to develop the sociology of technology, but a joint program applicable to both science and technology.

ST We will address it in a while. Before moving on, I would like to complete this background overview by discussing reflexivity, one of the four points of the Edinburgh Strong Programme, which was dismissed by the school of Bath, or at least by Harry Collins.

113. Kitcher 1994.
114. See in particular Rottenburg 2009.
115. See Pinch and Bijker 1984; Bijker, Hughes, and Pinch 1987.

1.5 Reflexivity

TP Reflexivity is an interesting topic as well. I have to tell you I have changed my positions on that. I wrote a piece called "Reflecting on Reflexivity" in the *EASST Newsletter*[116] in 1983. I was on a research fellowship in the Netherlands, and I wrote that at the same time I was doing "The Social Construction of Facts and Artifacts." "Reflecting on Reflexivity" was an antireflexivity piece. I was working with Bijker and Collins, and then I went to work with Mulkay and Ashmore, and the latter in particular convinced me that there was something in this reflexive project. We followed a degree of reflexivity ourselves, and I ended up publishing this paradoxical paper called "Reservations about Reflexivity and New Literary Forms: Or Why Let the Devil Have All the Good Tunes?" which is both pro and against because it uses a reflexive method to attack reflexivity, which is a nice paradoxical outcome.[117]

ST You also made a similar move a few years later—I mean, turning reflexivity against reflexivity—in the paper "Turn, Turn, and Turn Again,"[118] where you answered Woolgar's accusation that the field of social studies of technology was not reflexive enough[119] and that it applied the SSK framework to technology in an excessively formulaic way.[120]

TP Woolgar's ironic stance was getting on my nerves. In hindsight I realized how he might have seen that paper as being ad hominem, but I didn't mean it as ad hominem. I just found it strange that he would constantly attack an area for not being reflexive enough, making this reflexive move

116. Pinch 1983.
117. Pinch and Pinch 1988. In the paper, Pinch makes use of the so-called second-voice device to discuss new literary forms, finally—and paradoxically—dismissing them through the "main author's" voice as rhetorical, rather than reflexive, devices.
118. Pinch 1993b.
119. Woolgar 1991.
120. "I have nothing against a full-blooded reflexive version of the sociology of technology. Indeed, Woolgar's work elsewhere, and some of the positive suggestions for treating technology as text in his 1991 article, are important steps in developing such forms of analysis. What, for me, is a disappointment, is when the reflexive project on technology, like the reflexive project on science, can seem to get going only by first using the techniques of realist critique on the very people who open up technology for analysis" (Pinch 1993b, 519–520).

that was so easy to make every time. In fact this reflexive move was just a move, a particular sort of move.

Anyhow, the idea of reflexivity is that, since the sociology of scientific knowledge shows that knowledge itself is socially constructed, the question at some point must arise: "What about the knowledge that has been produced in the sociology of the scientific knowledge?" This is an old argument that was very important in the history of the field: because Robert Merton had first entered this debate many years ago, claiming the sociology of scientific knowledge was impossible precisely because of reflexive grounds[121]—you would be undermining the branch you were on. He had claimed it was an impossible thing to do. So anyone who works in this field is going to have to be aware of how to address the reflexive question at some point, and what follows from it.

I think my position, back in the days of writing that first anti-reflexive piece, was that deconstructing knowledge of yourself was not your job. No individual could do it, necessarily. An analyst coming from the outside could do it. And that was fine, but nothing followed for you as an individual analyst doing this work from that. You did not have to take it on board. I think that was the substance of my criticism, but then Ashmore, and Mulkay working with him, convinced me that we had a special responsibility to try and take on board this problem, and not just push it away, saying, "It's someone else's problem." The metaphor that I used in "Reservation of Reflexivity and Literary Forms" was the "the torch" (in Britain; in America it is the "flashlight" theory). You can shine a torch on somebody, and they can shine a torch on you, but what you cannot do is shine a torch on someone else *and* yourself at the same time. The answer is to replace your torch with a light bulb so that it can shine a light on everyone. Ashmore's idea was somehow that you can exhibit the reflexive potential in your own text at the same time as producing it. But what would that look like? There comes the issue.

Woolgar has written about this, distinguishing between different forms of reflexivity.[122] There is a sort of reflexivity which is psychological; it is

121. See Merton 1937.
122. Woolgar classifies

varieties of reflexivity along a continuum ranging from radical constitutive reflexivity to benign introspection. The constitutive reflexivity arises from [Garfinkel's suggestion] ... that, in any act of representation, there is an intimate interdependence between the surface appearance

when the individual is individually reflexive. This usually comes in a foot-note saying, "I was funded by so and so, I came into this field with these biases." That is like a psychological form of reflexivity in the history of the individual analyst. But there is a more interesting form of reflexivity, *constitutive reflexivity*, that aims to take on board reflexivity seriously. It says, "The knowledge that I am producing now as you are reading this text is itself socially produced, and I want to show you how that's done." So the very constitution of knowledge is a reflexive project, and you want to show that. But how do you do it? You have to try and disrupt the normal reading of the text. That was one of the approaches we were working with. To disrupt normal reading, one can start to put in, say, a different voice into the text. So you have to interrupt and ask yourself "Is this the authorial voice?" You know, this introduces a dialogue—that was a very common way of doing it. Mulkay wrote this play, *The Scientist Talks Back*:[123] a play on reflexivity that was about replication of experiments.

ST Somehow, the form of a paper is its content ...

(document) and the underlying reality (object). ... Representation and object are not distinct, they are intimately interconnected. By contrast, traditional conceptions of the natural sciences ... [emphasize this] distinction. ... As a result, apparent concessions to reflexivity, both in the natural sciences and in other disciplines which aspire to the Scientific Ethos, usually involve an entirely different form of reflexivity, which we call benign introspection. This kind of reflexiv-ity—perhaps more accurately designed "reflection"—entails loose injunctions to "think what we are doing." It is encouraged as a means of generating addenda to research reports, sometimes in the form of "fieldwork confessions," which provide the "inside story" on how the research was done. ... The social sciences fall awkwardly between constitutive reflexivity and benign intro-spection in virtue of their admission of some similarity relations [between document and object] and their pretentions to ideals of Scientific method. (Woolgar 1988b, 21–22)

For a more detailed taxonomy of varieties of reflexivity, see Lynch 2000.

123. Mulkay 1984, 265–266:

I decided to write a play, in the first place, because I wanted to bring the scientists, from whom we obtain our knowledge about science, into my secondary, sociological discourse as active par-ticipants. ... In this text, I have tried to present scientists' own words in a way which enables participants (with my help) to reply to and challenge some of the accounts of scientific action that have appeared in the recent sociological literature. ... A second concern which has influ-enced me in constructing this text is that of how to display the interpretative work carried out by analysts, including myself. ... In this text, I have used my empirical material to construct an imaginary dramatic confrontation. In this way, I have sought to emphasize the interpretative, creative character of my own, as well as of participants', discourse. ... Thus one of the things I am trying to do here is to explore seriously the issue of reflexivity in the sociology of science, and in sociology generally; and to find out whether new forms of analysis and presentation can help us to make our research practice appropriately reflexive.

TP Yeah, it is all that. It is a radical program. The only time I have ever done this successfully is in the article "Reservation of Reflexivity and Literary Forms," a fully reflexive article. I wrote it with a senior and a junior author. I thought it was successful, people liked that piece. It is very hard to do, but is actually easier when the object is the sociology of science. If the object is something else, like another science, this reflexive project is extraordinarily difficult. Basically, the program is not going anywhere because Mulkay has retired, Woolgar never managed to push it forward much further, and Ashmore hasn't either. It is very, very hard to do, and the only thing where we had any success was a book on health economics.[124] That was a book about the application of knowledge in the sociology of science to health: "Let's look at how economists apply their own knowledge to health, and then try and get some reflective tension over that, trying to make parallels between the two."

ST Even if its textual form is more conventional than the other paper of yours you quoted, I have really appreciated the reflexive effort of your paper on the testing of a clinical budgeting system—I mean "Technology, Testing, Text,"[125] where you deconstruct a previous version of the same paper[126] that, in turn, was aimed at deconstructing the testing of the system. This double deconstruction doesn't seem to me a self-referential game because, at the end the paper, it succeeds in accounting in a more articulated way the relationship between a technology and the different rhetorics used by specific social actors—sociologists included—to describe it and give it a meaning.[127]

124. Ashmore, Mulkay, and Pinch 1989.
125. Pinch, Ashmore, and Mulkay 1992.
126. Pinch, Ashmore, and Mulkay 1987.
127. Pinch, Ashmore, and Mulkay 1992, 285:

We have identified two broad forms of rhetoric: a strong-program rhetoric that draws on economic principle and carried the promise of radical change—change that can be tested and evaluated in an independent and scientific manner; and a weak-program rhetoric that is sensitive to the complex social and political realities of organizational change, presents clinical budgeting in an mild unthreatening way, views research on a clinical budgeting as a slow learning process, and recognizes that technologies are evaluated in a practical and political context.

[In the previous version of the paper] we established that a test in the natural-science sense of an experiment had occurred. We then deconstructed this supposed test. Finally, we ironicized the positive results that were obtained showing that they were "soft" in the sense of impressing important groups of people or arising from vaguely defined factors such as people "working better." ... What if we had taken the weak-program [of the rhetoric framing the object] as

TP That was part of the health economics project, it was about which rhetoric works in a particular context, I realized subsequently. I learned a lot from that project. I remember presenting that early work on budgeting to a bunch of doctors at a conference in Leeds: I pulled apart the arguments for this clinical budgeting, because these doctors were mostly opposed to it, and I showed them the logic of it. Then I showed them the logic of my own deconstruction: they hated that. But one guy came up to me and said: "That's brilliant!" Most of them were using these arguments in a practical context, and so a context of argument that most people are working in. I bet it seems pointless if you just ignore that. Sure, one person will find it brilliant, but the rest of them would ask, "Why the hell did you do that?"

ST And that is how you took on board reflexivity after your initial resistance. Actually, in your paper[128] coauthored with yourself, you talk about— or rather, you give "proofs" of—the existence of a previous, unpublished paper where you took your distance from Collins's aversion toward reflexivity, advocating the adoption of a more reflexive stance in SSK. I am half-tempted to ask you if it really exists, but I think it is probably better to leave the issue unsolved.

---

definitive? What would our deconstruction have shown then? The answer is: not very much. ... The difference between our rhetoric and that of the participants is that for our textual purposes we brought the two rhetorics into opposition to achieve "mutually assured deconstruction"; on the other hand, the participants seem to have used the two rhetorics either separately or side by side to reinforce their arguments. They have managed to avoid the potential deconstruction that the use of the dual rhetoric can entail. (ibid., 280–282)

128. See Pinch and Pinch 1988.

2 SSK at Large

2.1 The *Golem* Trilogy

Simone Tosoni The 1980s was the decade in which SSK and, later, STS became solidly established academic disciplines, after their pioneering phases—notwithstanding significant opposition and sometimes open hostility, especially from realist philosophers, for the relativistic stance we have already discussed. Up to the 1980s, anyhow, that debate had been mainly academic, involving a very specialized audience. In the 1990s, a new front opened: the debate over scientific knowledge as a social construct reached a wider audience, becoming public. A relevant role in this process was played by you and Collins: I am thinking of course about your book, *The Golem*,[1] published in 1993, which popularizes the main acquisitions of your approach to scientific controversies. Or, rather, I am referring to the whole *Golem* trilogy, since the first volume on science was followed five years later by a second volume focusing on technology, *The Golem at Large*,[2] and more recently, in 2005, by a third one on medicine, *Dr. Golem*.[3]

Trevor Pinch That is true, but it was not our intention. The project took shape while it was going. You need to know the history of that book. I'm gonna tell you the background to the book. The idea for *The Golem* was originally started by Collins with his friend, Steven Shapin, the historian of science who is now at Harvard. They were kicking around the idea of redoing something called *The Harvard Case Histories in the Experimental*

1. Collins and Pinch 1993.
2. Collins and Pinch 1998.
3. Collins and Pinch 2005.

Science,[4] which is a classic case studies collection edited by James Bryant Conant, former president of Harvard and a very eminent historian. The original idea was to apply the ideas of the sociology of science to these classic case studies. Collins started to work on it with Shapin, but somehow Shapin dropped out of the project (I don't know why, you'd have to ask Collins), so Collins started looking for another coauthor. His idea, originally, was that this book would be mainly aimed at a science education audience. At first he approached a colleague of mine at York, Robin Millar,[5] who had been trained at the Edinburgh Science Studies Program with Barnes and Bloor. Millar was working on the implications of SSK, the sociology of scientific knowledge, for the world of science education; he was trying to figure out how to explain it in the classroom because, maybe, if we talked more about the process of science in the classroom, science education would actually improve. There was a lot of interest in such a program back then, and there still is. I often get asked about that even today. When I give lectures, this is one of the questions I always get: what are the implications for science education? So Collins asked Millar, but Millar was too busy. He was at York, I was at York, I had been Collins's previous collaborator, and he was my mentor. So, after Millar turned him down, since I was at York and I knew about the project from Millar, Collins said, "What about you, Trev? Would you be interested in editing this *Golem* volume with me?" The question was that we knew that this book would only work if it reached a popular audience, but Collins had originally envisioned it for science education people. When Millar dropped out, it was clear that we had no credentials in science education, and so we could not write that sort of book. Well, let's try writing and see if we can just write it for a general audience, and see what it would look like, we thought. This was almost an experiment. So I wrote a chapter. You know Collins, this is the way he works: "Why don't you try and write a chapter and see if you can do it. Choose the easiest thing you've got: solar

4. Conant 1948. This edited collection aims at fostering a better public understanding of the scientific method and practices (in a realistic perspective) and is "designed primarily for students majoring in the humanities or the social sciences [since they] require an understanding of science that will help them to relate developments in the natural sciences to those in the other fields of human activity" (vii).
5. Professor Robin Millar is a member of the department of education of the University of York since 1982, and he has extensively published on science education, public understanding of science, and scientific literacy.

neutrinos." So I wrote it up, sent it to him, and said, "This is what a popular account of solar neutrinos would look like." And he loved it: "This is really great!" He still thinks this is one of the best bits of writing in the whole *Golem*; he thought it was perfect.

So now we were thinking this was going to be a book for lay people about the nature of science, and so the audience—the idea of what the audience was—slowly evolved. It switched to the general public. We had the whole metaphor of the golem. The case studies we had were mainly our own case studies because they were easier to use. Also, we wanted to look at the sociology of science from the Bath School perspective, this Empirical Programme of Relativism. We picked our own case studies. Collins did work on Einstein, and there was a guy who did the stuff on the lizards that I wrote up in *The Golem*, the lesbian lizards! He was Greg Myers, from Lancaster. He has vanished from the field now.[6] And then Travis's case on memory transfer,[7] and a bunch of people we knew all became "golemized." The book would come out as a hardback with Cambridge.

There was a lot of fighting over the title because Cambridge, actually the New York office of Cambridge University Press hated the word "golem." "You can't call a book *Golem*, no one knows what a golem is!" Collins was Jewish, so he came up with the golem metaphor, since in his family there was this tradition of a golem being like a big gardener, sort of a friendly, big guy. It was a bumbling giant, and that was exactly the golem metaphor we were looking for. So the guy at Cambridge University Press in New York did a survey on his office corridor, and no one knew what a golem was. He told us, "Here we are: Jewish, New York, no one knows what a golem is. What chances can a book with that title get? Why don't you call it something like *What Everyone Should Know about Science*? Have that as the main title?" We stuck with *Golem* and answered, "No, that's boring. We like this metaphor, we think that we're onto something. It's a good metaphor." It seems we were very prescient because the metaphor of the golem became popular a bit later. We started to write the book when I was at York, and we finished it in 1991 or 1992. At that time the Internet was just getting going, and this helped to spread the "golem" metaphor. But it's true that at the beginning, no one knew what a golem was. It wasn't popular at all. Then, suddenly,

6. Myers 1990.
7. Travis 1987.

Pokémon cards came out with golems, and there were golems in various online games, and suddenly there was that book, the story of the original golem, the monster of Prague, which won a literary prize,[8] and suddenly golems were everywhere. Then everyone knew what it was, and we were very lucky that we rode that wave. It helped to make the book famous. The book won various prizes, and it is currently out with Canto Classics, the popular wing of Cambridge, in paperback. They are redoing all their Canto books now, and as there are just a few "Classics," it's becoming a sort of famous book. Eric Hobsbawn's book, *The Invention of Tradition*, is in Canto Classics;[9] only a few of these books are going to be in this now. It is a good seller, and at Cambridge they were delighted. It will probably stay in print for a very long time, which is great.

ST   So, why a golem? How can science, technology, and medicine be compared to a golem?

TP   As we say in the introduction of the book, there are very few metaphors for describing science and the activity of scientists. One of those common metaphors is the scientist as being godlike, but that invests science with too much certainty. The other very predominant metaphor was Frankenstein from the marvelous novel by Mary Shelley where Frankenstein is the scientist, but the *monster* becomes associated with the name Frankenstein. That's a very popular metaphor for everything wrong with science. We wanted to escape those two views of science, which we saw as "the public flip-flopping" (a term we took from transistor technology, where you have flip-flop devices). People would flip from the model of scientists being gods, to the other model: "Oh, the scientists are totally corrupt and evil," you know, "in the pay of capitalists; there's nothing in science." So we wanted a middle way between those two. The golem is nice because there are a lot of things in it. There is the natural world: it is made of clay, so it's coming from the natural world. And it's about incantation, history, and culture to some extent (Jewish culture, in this particular case). So the golem is a form of knowledge that would protect mankind if we only understood. It was a little bit stupid, like a bumbling giant of science, but once you understand

---

8. In 1997 David Wisniewski's *Golem* (New York: Clarion Books, 1996) won the Randolph Caldecott Medal, awarded to the "most distinguished American picture book for children" published in the preceding year.

9. The classic by Hobsbawm and Ranger was reprinted by Canto Classics in 2012.

what it is, it stops being frightening but remains powerful. That was the whole idea.

ST   And it has "truth" written on its forehead.

TP   Yes, in Hebrew. Very evocative. So once we've got that metaphor out there—and it seemed to be having success—then, obviously, when we wrote the next book on technology,[10] we went for it again. That second book, the one on technology, was my idea. One of my grad students, Park Doing, who plays with me in the Atomic Forces,[11] actually came up with the subtitle "Golem at Large." We were messing around with different titles: *Golem on the Loose*, you know. We wanted something catchy: the golem moving into the world once more. So *The Golem at Large* became a technology book, and then the medicine one,[12] *Dr. Golem*, just seemed a funny title.

ST   Will it be the last of *The Golem* saga?

TP   We are writing a fourth one, a teaching *Golem*. It is a very unusual *Golem*, it is going to be called *The Companion Golem*. It will feature entries for all the key concepts of the sociology of science. We want it to be useful for teaching because we have discovered over the years that no textbook properly explains the ideas of the sociology of science. Collins and I are constantly lecturing about the nature of the sociology of science to a wider audience and also to scientists, so we thought that there could be room for a book that just explains some of these basic ideas that we have found to be useful. It will be an unusual book because, as we started to write, Collins and I immediately started to disagree over the key ideas and why we thought they were important. This stems from a rather bitter disagreement we had in *Dr. Golem* about the vaccination issue, at this very table. The key point is that he and I, for the first time, had a huge blow up, I think because medicine is a much more personal subject. It is amazing when you think

10. Collins and Pinch 1998.
11. Starting in 2009, Trevor Pinch has collaborated with several bands as a performing musician, playing his self-made synthesizer (see Pinch 2007). He is a member of the Atomic Forces (Ithaca, NY) (Folk Machina, online release 2014, https://theatomicforces.bandcamp.com/releases, last accessed Apr. 6, 2016) and, with James Spitznagel, of the electronic experimental band Electric Golem (*The Electric Golem*, Ricochet Dream, 2010; *Sky Snails*, Periphery, 2011; *We Are the Acid*, Level Green Recording Company, 2012; *Smiling Like an Angry Turtle*, Level Green Recording Company, 2013; *Astrogolem*, Level Green Recording Company, 2015).
12. Collins and Pinch 2005.

about it: of all the experts you have talked to in your life, the consultation with the doctor is the most consequential one because a doctor can literally tell you you're going to die in a few months' time. Nobody else, not your tax inspector, not the plumber, none of these scientists, none of them is more consequential other than, you know, perhaps a judge sentencing you to death. So it becomes a much more personal issue, and medicine has more uncertainties, so we started fighting over some of the implications of the sociology of scientific knowledge applied to medicine. We disagreed about the role of parental expertise: how far can it go?[13] I thought it could go further than Collins, and we had this bitter fight.

This antagonistic spirit has entered somewhat into this last *Golem*, and at one point we couldn't solve our issues—not over medicine, but over the sociology of science. For instance, I think "inscription device"[14] is an important idea in the sociology of scientific knowledge: Latour has developed this. He talks about it as a device for translating from the "material world" into an inscription that can be graphically represented, that is reproducible, and that can be sent, for example, with a fax machine, or that can be copied by other people. So it becomes, in his terms, an "immutable mobile"[15] that can be moved around scientific networks. An inscription is something like a chart recorder, a bioassay. I find it an important idea; Collins doesn't. The disagreement with Collins is—I think—based on the notion that the inscription, once inscribed, has power. Collins would say, "No, it's always an interpretive context. Inscriptions themselves say nothing about the interpretive context. It's about the way you put it in the process." He means that anything that takes that away from interpretation is bad. That is the substance of our disagreement. So how are we going to resolve it? I cannot

13. For a further discussion, see section 3.6.
14. See chapter 1, note 90.
15. Latour 1986, 6:

The essential characteristics of inscriptions cannot be defined in terms of visualization, print, and writing. In other words, it is not perception which is at stake in this problem of visualization and cognition. New inscriptions, and new ways of perceiving them, are the results of something deeper. If you wish to go out of your way and come back heavily equipped so as to force others to go out of their ways, the main problem to solve is that of mobilization. You have to go and to come back with the "things" if your moves are not to be wasted. But the "things" have to be able to withstand the return trip without withering away. Further requirements: the "things" you gathered and displaced have to be presentable all at once to those you want to convince and who did not go there. In sum, you have to invent objects which have the properties of being mobile but also immutable, presentable, readable and combinable with one another.

imagine a companion book without a reference point, or an entry if you like, of two or three pages explaining what an inscription device is and how it works. Probably I will write it up, and then Collins will write his reply. So we will have double entries when needed, where we argue as to why this is a good or bad idea. It's going to be a strange teaching book, one with a reflexive level. A metalevel.

Collins also tried to get other *Golem*s done. He has approached others to do one on artificial intelligence and robotics, but none of them have come out. I would like to do one on fringe science: I think it will be perfect. Collins is uneasy about fringe science because, as we said, once you start looking at fringe science, you can get tied with it; stigmatized, somehow, for supporting it. But I think it would be a real fun book to do. Anyway, we are going to do just this "companion"—when we get the time ...

ST   Good luck with that!

2.2   Science Wars

ST   I think we should introduce your work on the social construction of technology before we discuss the *Golem at Large* and *Dr. Golem*. For now, I would like to ask you about the first *Golem*, on science, and in particular about the role it played in the so-called science wars.[16] I am thinking, of

16. As Collins summarizes on his personal webpage on the Cardiff University website (http://www.cf.ac.uk/socsi/contactsandpeople/harrycollins/science-wars.html, last accessed Apr. 6, 2016):

The "science wars" began in the early 1990s with attacks by natural scientists or ex-natural scientists who had assumed the role of spokespersons for science. The subject of the attacks was the analysis of science coming out of literary studies and the social sciences. ... What has attracted the ire of the critics is the turn toward the social analysis of the content of science in addition to analysis of its social organisation. A new field, now known as "the sociology of scientific knowledge," or SSK, began in the 1970s. One of its principles was the tenet of "symmetry" which means analysing the scientifically true and the scientifically false in the same way. When it is first encountered, scientists find this an uncomfortable perspective; in the way it ignores scientific truth, and the way it tries to set the conclusions of science in a social perspective, SSK can seem like an attack on those things that scientists hold most dear. ... Attacks by science warriors often take on the characteristics of a "witch hunt" instead of an academic debate. For example, "relativism"—a subtle philosophical idea with a number of meanings—is sometimes treated as synonymous with "anti-science." An accusation of relativism is taken as sufficient in itself to render further argument unnecessary. And the arguments and political tactics adopted by the science warriors seem less designed to convince their academic opponents of their errors than to convince an outside audience.

For a detailed account of the science wars, see Segerstråle 2000.

course, about the way the book aims to reach a wider, and popular, audience. It is a fact that if you delete the word "truth" written on the forehead of the golem—actually, you delete a letter—you destroy it.

TP I think that the science wars had been already building up, but *The Golem* definitely did have a role in it. David Mermin, a Cornell physicist and a colleague of mine, reviewed *The Golem* in two columns of the journal *Physics Today*,[17] a mainstream American Physical Society publication—he devoted a lot of attention to it. But *The Golem* itself doesn't figure that much in the science wars, which basically were initiated by the well-known book by Gross and Levitt, *Higher Superstition*.[18] And then along came Alan Sokal and the Sokal hoax.[19] The people they are picking upon more are feminists like Sandra Harding[20] than social constructivists of science and knowledge. *The Golem* does not attract much attention in Gross and Levitt, and in Sokal. By and large, my work with Collins was not really taken on board. But Collins then got in a fight, at the British Association for the

17. Mermin 1996. While appreciating *The Golem*'s purpose to demystify the myth of the scientist as a godlike figure ("the scientist as a sorcerer"), Mermin firmly rejects its methodological relativism as an inadequate approach to account for science in the making: "The pertinent issue in assessing the claims of The Golem is not whether scientific truth is determined by constraints from nature or from social construction but whether Collins and Pinch strike a satisfactory balance between these two aspects of the process. I think their book furnishes an instructive demonstration of what can go wrong if you focus too strongly on the social perspective. By paying insufficient attention to how nature does constrain us, Collins and Pinch draw lessons about the building of scientific consensus that leave out a relevant part of that process" (Mermin 1996, 11).

18. Gross and Levitt 1994. The bestseller *Higher Superstition* is a harsh attack against the postmodernist "academic left" (in particular, scholars in cultural studies, literary theory, and science studies), accusing them of advocating an antiscientific standpoint while lacking any real, in-depth awareness of scientific culture.

19. The "Sokal hoax" is discussed later in this section. For an account by the author, see Sokal 2010.

20. Sandra G. Harding is a philosopher of science, a postcolonial theorist and a feminist epistemologist. During the "science wars," she was vehemently attacked for her refusal to acknowledge science as universal and neutral and for her depiction of the scientific enterprise as Western-centered and androcentric. See in particular Harding 1986, 1998, and her recent work, *The Postcolonial Science and Technology Studies Reader* (2011).

Advancement of Science, with a leading British scientist, Lewis Wolpert,[21] who was involved in the science wars. Wolpert was reading works by Andy Pickering out of context, and Collins actually accused him of being a liar, and they had a huge fight. I wouldn't say that it was *The Golem* itself that had led to the science wars, though. It may have played a small role in a bigger general context.

If it was anything, it was that the ideas in the sociology of scientific knowledge were getting more and more attention, and the person whose work was getting the most attention was Latour. When I first came to Cornell in 1990, I remember some guy coming up to me—I can't remember who—and saying, "Oh! You do what Bruno Latour does." And I said, "No, *he* does what I do!" because we had been in this field longer than Latour, although he became much more prominent and well known. He did bring a lot of attention to it because he wrote about Einstein's work.[22] Mermin, for example, was outraged by Latour's paper that claimed that relativity was somehow a social process. Funnily enough, after hearing Latour lecture, Mermin started to defend him against some of the wilder attacks from his physicist friends.[23]

I'll tell you, there is just one specific incident I know where *The Golem* got up someone's nose. A prominent physicist at Cornell, in a very angry mood, came to find me when I was acting chair of the Department of Science and Technology Studies. He stormed into my office with *The Golem* open: "How could you write this?" He was outraged! He obviously interpreted *The Golem* as an attack on science and on physics in particular. I was very calm about it. I simply pointed out that he was misinterpreting, and so I patched up the quarrel. It was just typical for a physicist. What had happened was that this very famous physicist had a friend who was a philosopher of science, and I had debated some of those issues with this philosopher a few years earlier. The philosopher did not share my positions,

21. Lewis Wolpert is a British biologist and a very well-known science popularizer. His quarrel with Collins at the British Association for the Advancement of Science at Loughborough in September 1994 is one of the most widely known "showdowns" in the science wars. As Segerstråle euphemistically recalls, "at that conference, it became apparent to many that social constructivists and scientists had difficulty speaking to one another" (Segerstråle 2000, 7).
22. Latour 1988a.
23. See Mermin 1996, 1997.

but he thought my work was serious and important, so he told the physicist, "No, you got this wrong." I think this physicist thought I was some scurrilous person, attacking science with no credentials at all, but the philosopher told him, "Actually, this guy is a good guy. His work is good, you should take him seriously." So the physicist came to apologize, and he's been apologetic ever since, kind of, "Oh, yeah, I've actually talked with a philosopher, and I realized I misinterpreted what you have written, and you're not beating up on scientists after all."

ST Still, the fact that many leading scientists and physicists decided to debate with you means that, in the 1990s, SSK had definitely become socially visible.

TP Yes, suddenly we were socially visible, and the leading physicists knew of our work, but I don't think it was just because of *The Golem*. I think the field of the sociology of scientific knowledge had come to more prominence. Before that, it was a marginal field, with figures like Latour, maybe our books as well, and the feminists' work on science, particularly Donna Haraway. Suddenly there was something in the air, suddenly people were aware of us: "Hey, there's a group of people out there saying that science is a social construct. Isn't that crazy? How could they be saying that?" Then the Sokal hoax was the thing that really brought it out; it was a huge thing. It made the front page of the *New York Times*.[24]

ST I guess you had already met Sokal at some point before.

TP Oh, yes, it was funny. I met Alan Sokal subsequently and debated with him and his main coauthor, Jean Bricmont. They did a follow-up book.[25] As a sociologist of science, you have to know about the Sokal hoax. He basically wrote a fictional paper[26] for a journal called *Social Text*, which wasn't a peer-reviewed journal. You know, there are some people in our field who draw much more upon postmodernism and its idiom; he went to talk with Donna Haraway and her students, I think he got a lot of their language, and in his paper he was deconstructing gravity using Derrida, moving toward literary criticism and stuff like this. He poured into it a mixture of ideas from postmodern thinking, Deleuzian ideas as well, and he mixed them all up together in a sort of incoherent way. Somehow, he got it published. It

24. Scott 1996.
25. Sokal and Bricmont 1997.
26. Sokal 1996.

would have been more powerful if he got away with it in a peer-reviewed journal! I still think that Alan Sokal's original paper would never have been published in *Social Studies of Science*, but *Social Text* wasn't a proper peer-reviewed journal. However, he had a point, because it was very funny. If people had read it with more humor, rather than getting so uptight, maybe he wouldn't have got into this huge debate about it. People got really angry at Sokal instead of going, "Actually, it's pretty funny." The editor of *Social Text* was Andrew Ross, who has never been in the sociology of science, ever. I met him once when he first came to Cornell; he was lecturing at the Society of Humanities, but he is not a sociologist of scientific knowledge at all. Did you read his book, *Strange Weather*?[27] He is much more a Marxist critic, a literary guy or a cultural studies person, than a sociologist or historian of science. Anyway, the paper drew dramatic attention and got on the front page of the *New York Times*. Then Stanley Fish got into the debate in an early letter, comparing science to baseball.[28] The whole thing got a bit out of hand, especially in America.

27. See Ross 1991.

28. See Fish 1996:

Why then does Professor Sokal attack [the sociologists of science]? The answer lies in two misunderstandings. First, Professor Sokal takes "socially constructed" to mean "not real," whereas for workers in the field "socially constructed" is a compliment paid to a fact or a procedure that has emerged from the welter of disciplinary competition into a real and productive life where it can be cited, invoked and perhaps challenged. It is no contradiction to say that something is socially constructed and also real. Perhaps a humble example from the world of baseball will help make the point. Consider the following little catechism: Are there balls and strikes in the world? Yes. Are there balls and strikes in nature (if by nature you understand physical reality independent of human actors)? No. Are balls and strikes socially constructed? Yes. Are balls and strikes real? Yes. Do some people get $3.5 million either for producing balls and strikes or for preventing their production? Yes. So balls and strikes are both socially constructed and real, socially constructed and consequential. The facts about ball and strikes are also real but they can change, as they would, for example, if baseball's rule makers were to vote tomorrow that from now on it's four strikes and you're out. But that's just the point, someone might object. "Sure the facts of baseball, a human institution that didn't exist until the 19th century, are socially constructed. But scientists are concerned with facts that were there before anyone looked through a microscope. And besides, even if scientific accounts of facts can change, they don't change by majority vote." This appears to make sense, but the distinction between baseball and science is not finally so firm. On the baseball side, the social construction of the game assumes and depends on a set of established scientific facts. That is why the pitcher's mound is not 400 feet from the plate. Both the shape in which we have the game and the shapes in which we couldn't have it are strongly related to the world's properties. ... Even if two activities are alike social constructions, if you want to take the measure of either, it is the differences you must keep in mind. This is what Professor Sokal does not do, and this is his second mistake. He thinks that the sociology of science is in competition with mainstream science—wants either to replace it or debunk it—and he doesn't understand that it is a distinct enterprise, with objects of study, criteria, procedures and goals all of its own.

ST   When did this climate of hostility reach its peak? How are things now?

TP   It became harder in this period of *The Golem*. I can remember actually getting my first-ever anonymous hate mail when I was teaching here
in America after *The Golem* was out. I never had these before, you know.
Stitched … one of those letters with bits of text put together, which I
handed in to the police. It was strange. I can't remember where it came
from. It was in my mailbox, and basically it said that nasty things would
happen to me if I wrote another book like *The Golem*.

ST   That's incredible!

TP   Yeah. Somebody was really offended. I took it to the police, but in
hindsight I don't think it was such a big thing. Sandra Harding, whom I
met at a conference around then and is now a friend, told me she got loads
of these letters. I only got this one. It was badly done; they had cut out
bits of text to hide their own handwriting. But it was serious enough to
convince my wife and me to delist our telephone number and adopt a low
profile, because it was a death threat: "If you publish another *Golem*, you
know we're gonna kill you," or something like this.

ST   It would have been great to have it as the cover of this interview book!

TP   The police certainly saved it, we can probably get it back. Actually, I
don't know why I didn't photocopy it at that time! I think this shows that
I took it really seriously; I didn't think it was something I could use! I was
sort of scared. The police looked at it carefully because it is a very serious
offense, especially as there was a postmark on it, which made it worse.
American law does not tolerate such frightening messages passing through
the US mail. It is a really heavy-duty thing, a federal offense; you could be
put away in prison for a very long time. And at that time there was that
paranoia produced by the Unabomber.[29] He was still out there. The FBI

29. Theodore John Kaczynski (1942), also known as the Unabomber, was arrested
in 1996 for killing three people and severely injuring more than twenty with self-
made bombs, frequently sent to his victims through the mail. On September 19,
1995, at the FBI's request, the *New York Times* and the *Washington Post* published his
manifesto, "Industrial Society and its Future," a call for a global revolution against
technological society. In his pamphlet, Kaczynski revealed a certain awareness of
the philosophy of science and the sociology of scientific knowledge, seen as typical
examples of the "craziness" of "Leftism":

Modern leftish philosophers tend to dismiss reason, science, objective reality and to insist
that everything is culturally relative. It is true that one can ask serious questions about the

had shown up at a Society for Social Studies of Science conference, looking for the Unabomber hanging out with us, because the Unabomber had written those antiscience and technology pieces. They thought there was some possibility that somebody in our field could somehow be linked to the Unabomber, or there could be connections; they were sniffing around and showing up at our meetings, standing out in their suits. They also interviewed people in our field, you know, and wanted to know if anybody knew anything about the Unabomber. It was the same period, in the 1990s.

You know, there were also a few denial-of-tenure cases in America where there were worries over the science war issue. It happened first to Bruno Latour: he had been offered a position at the Institute for Advanced Studies at Princeton, but he had been turned down by scientists who couldn't stand him. And then there was Norton Wise, a much milder historian, who was beaten up by a very famous Nobel Prize winner, a physicist who publicly attacked him. Wise was offered and then denied a post also at the Institute for Advanced Studies[30] and then went to UCLA. That's quite a public bashing he got ... these were the casualties of the science wars. There were worries that people had been turned down for tenure in America because of the sociology of science.

---

foundations of scientific knowledge and about how, if at all, the concept of objective reality can be defined. But it is obvious that modern leftish philosophers are not simply cool-headed logicians systematically analyzing the foundations of knowledge. They are deeply involved emotionally in their attack on truth and reality. They attack these concepts because of their own psychological needs. For one thing, their attack is an outlet for hostility, and, to the extent that it is successful, it satisfies the drive for power. More importantly, the leftist hates science and rationality because they classify certain beliefs as true (i.e., successful, superior) and other beliefs as false (i.e., failed, inferior). The leftist's feelings of inferiority run so deep that he cannot tolerate any classification of some things as successful or superior and other things as failed or inferior.

30. The *Economist* ("You Can't Follow the Science Wars without a Battle Map," Dec. 13, 1997) describes the incident as follows:

Unhappily, the science wars are taking casualties. A notorious case is Norton Wise, a historian of science at Princeton University who applied for a post at the Institute for Advanced Study. In May, after a panel of three institute and three outside experts voted 4 to 2 in his favour, the director of the institute decided against his appointment. Mr Wise's opponents noted not only his relatively sparse publishing record but, perhaps more pertinently, a letter from him to the New York Review of Books rebuking Steven Weinberg, a physics Nobelist, for his enthusiastic praise of the Sokal hoax (in an article which had irked many other moderate science-studies scholars). The letter, it appears, was taken as a sign that Mr Wise had strayed from orthodoxy.

In 1996 Wise had critically reviewed Gross and Levitt's *Higher Superstition* for the journal *Isis* (Wise 1996), becoming de facto involved in the science wars.

But there is another point in all this: David Mermin and I had these sort of fairly friendly but pointed exchanges. Now, in my class I teach at Cornell, What Is Science?, I often bring in scientists to interview. After George Bush became president, I brought Mermin back in the class to talk with him about the science wars, to talk over some of those issues. This was two or three years into George Bush's presidency, so it was after the Iraqi invasion, probably around 2005. Mermin said he just couldn't understand why he got so excited about the science wars, why he thought it was worth investing in reviewing books like *The Golem*, because something far more serious was happening: Bush's attacks on science and the rise of Christian fundamentalism with Bush. The scientists were much more concerned about this. In a way, this was good for the sociologists of science because it made the scientists focus on the real enemies—who really were antiscience, had enormous power, and were reducing their funding and attacking the institutions of science—compared with this sort of petty annoyance caused by these few people who were social constructivists. That put it all in perspective.

ST What is the situation like at the moment?

TP The situation today is that the Interdisciplinary Science Studies Department at Cornell, established in 1992 by Sheila Jasanoff, is flourishing as far as I can see. One of the keys to the success at Cornell was the integration we managed between historians of science and technology and sociologists. There has been a growing problem in universities as a whole because of the stock market crash in 2008 and the lack of resources going around. But in terms of the field, the success of Cornell's Science, Technology and Society program is also a hallmark, a mark of how well it is doing. Under Sheila's guidance we really built up an excellent graduate program—thinking seriously about how to train graduates in this new field—and many of our graduates have gone on to make their own mark in the field. Another sign of success comes from the NRC (National Research Council), which rates graduate programs in America every five years in a systematic way. In their last ranking, they claimed there were only two new emerging disciplines in the American academy that they were going to look at and that were of serious enough concern to be ranked; one of these is the social studies of science, or as they call them, "science studies." "Science and Technology Studies," I think, was the actual denomination. So it is an emerging new discipline, a new field. This week I am going to visit a college to review their STS program. Mike Lynch, Ron Kline and I are spending a lot of time going around the country, reviewing STS programs that are expanding. Our graduate students are getting jobs, and the appeal seems to be growing.

ST   Now that we can consider the science wars definitely over and the field firmly established: what have you learned from these debates with scientists? How have these quarrels contributed to the development of the field? Did you get anything from them?

TP   Great question. Let me answer in two different ways. First: the program at Cornell used to be a "Science, Technology and Society" program, and it was originally set up in such a way by scientists who, as a category, have always been involved in our program. Here's my feeling about this, just an intuition as to why this is important in the science wars: if this field remains just as a humanities field, it may not go very far, it may wither. Actually, in the long run, the scientists are going to be one of the most important audiences for the field. Therefore, having them understand what it is about and dispelling any hostility is a very important thing. I believe that the sort of work we do, technically engaged with the scientists, is actually something of interest to the scientists. They actually often show up at our talks, and we are still recruiting some science students who are interested in this area.

So one thing I think I have learned and reinforced in my mind is the importance of having scientists as an audience: we don't want to lose them. You know, we are doing empirical studies, theoretically informed, within a sort of canon of—however you want to call it—empiricism; that isn't so far from what they're doing, compared to someone more in the humanities, who has a different set of theoretical concerns. So I'd be very worried if we didn't have scientists on board. That is why I'm always teaching scientists, and I like it very much, they are one of the most interesting audiences. It is a challenge, of course, but once they get the whole thing, some of them really come on board. They see what we are trying to do. It is certainly one of the major things I have learned.

What else have I learned? Well, it really surprised me that I could get frightening hate mails for writing a book like *The Golem*. That was more about the extremes, maybe, of American culture. The idea that writing an academic book on anything could generate hate mail is surprising; but a book about something so obscure as golem science? It is not like, you know, writing about religion or sex, or anything you might think people could get really upset about. So I really can say I learned something from that, which was just a surprise.

ST What about the methodological and theoretical point of view?

TP Yes, the other thing I was mentioning that I learned from the Science Peace Conference[31] is how extraordinary difficult it is to make any progress in debating these issues. The amount of stuff, baggage, that you have is overwhelming. This is like something we all know when you write something: you actually know you're right. But when you enter into argument, you have far more baggage with you about things you just assume about the world, how an argument works, what counts in an argument, and the sort of evidence you have. When you sit down face-to-face with scientists and debate this, as we are trying to do now, you suddenly realize, "My God, there is a huge chasm here." That's why we had these rounds in the Science Peace Conference: we tried to overcome this by circulating each side's papers again and again.

ST The book's structure is very interesting because it is structured in a very similar way.[32]

31. The so-called Science Peace Conference was held at the University of Southampton in July 1997 and is recalled by Labinger and Collins in the preface of *The One Culture? A Conversation about Science* (Labinger and Collins 2001), the volume that represents its ideal prosecution:

In May 1997 Mike Nauenberg, a physicist at the University of California at Santa Cruz, organized a small conference which brought together some of the critics and the criticized [in the science wars]. There, Collins was able to meet Alan Sokal and to continue his discussions with David Mermin. Interestingly, in the setting of that conference, Mermin and Collins often found themselves pushed together by common disagreement with Sokal. Around the same time, Collins, then at Southampton University, was organizing the so-called Southampton Peace Workshop, which took place in July 1997. Sokal was unable to come, but the participants included Labinger, Mermin, and Pinch, as well as several other scholars representing diverse fields: physics, history of science, literary theory. For two days, one of which was spent cruising round Southampton water in a small motorboat, the eight conferees were closeted together for intense discussions. Only on the third day, after some mutual trust and understanding had been built up, did the workshop become a public event. This contrasted with other science wars forums, where opposition was usually absent and it often seemed that a public display of scorn was the chief purpose. (x)

32. "The particular format we have chosen for the book tries to reflect the successful give-and-take structure of the Southampton workshop. Initially twelve authors (some with coauthors), drawn equally from the ranks of science studiers and practicing scientists, contributed position papers on a theme of their choice appropriate

TP Yes, to try to get people to read more carefully, try and see if there is any change. At some points I was starting to see an emerging understanding. I think that some understanding emerged with physicists like David Mermin, who were really engaged and sensible. Regarding Sokal, by the way, I don't take him that seriously. I have met him a few times now. I think he just found a brilliant career move. He was a kind of failing physicist who wrote a very funny paper that made him world famous. He had a gig, suddenly. I don't think he was serious about trying to understand the field. Talking to him in private, you know, he never really got a deep understanding of what anyone in our field was doing. I can tell you, this is a physicist who took up a bit of lingo. There was also this Bricmont, a crazy Belgian guy he worked with: he wasn't serious at all, I realized. I mean, Sokal is serious, he is just not very good. So, in hindsight, this explains what I learned from it.

---

to the overall topic. The papers were then distributed, and everyone was invited to comment on whatever aspects of the papers evoked their interest or criticism. In the third and final round, authors were given the opportunity to respond to these commentaries and to defend, clarify, or even modify their positions" (ibid., 9).

·

3 Social Construction of Technology

3.1 An Integrated Program for Science and Technology

Simone Tosoni   As we have seen, more or less at the same time that your influential paper on externalities in scientific observation was published,[1] you had another groundbreaking paper out, "The Social Construction of Facts and Artifacts," coauthored with Wiebe Bijker,[2] which contributed so significantly in opening the new field of the sociology of technology.

Trevor Pinch   In that paper we were looking for a common language to analyze both scientific and technological controversies. Since technology was the new thing at that time, it so happened that most of the emphasis was upon technology, but it was actually meant to be equal.[3]

ST   You explicitly make this point in the paper; that is why the first part of the interview focused in such detail on your early work in SSK. Actually, what strikes me most is that we could say that you and Collins had already developed an approach to technology in that early work. When dealing with an experimental apparatus, the process of production of scientific

1. Pinch 1985b.
2. Pinch and Bijker 1984.
3. Pinch and Bijker 1986, 349:

In brief, our argument for treating science and technology within the same framework stem[s] from three considerations: (1) the unsatisfactory nature of the attempts to demarcate between science and technology; (2) the practical problems encountered by researchers investigating the science/technology relationship, and, in particular, the difficulties of distinguishing the separate contributions made by science and technology to particular innovations; and (3) the concrete demonstration (with examples drawn from our own empirical work on science and technology) that problems and issues raised by the study of science are similar to those raised by the study of technology.

knowledge is actually the same process through which the characteristics of that technological system are defined.

TP   Absolutely. What I did (and Latour did) about the "black boxing" of science was really about making it more technological, in a way. So there were many parallels, although many people back then made a distinction: they thought of technologies intuitively as being not bits of scientific apparatus, but things like bicycles, missiles, things that look somehow different. Lightbulbs or materials like Bakelite.

ST   Whereas in your case studies, when you describe an experiment, a hypothesis on nature is at the same time also a hypothesis on the behavior of a piece of technology. It can be an argon trap, or a gravitational wave detector antenna, but the controversy over the existence of a natural entity is, at the same time, a controversy on how a technological artifact is supposed to work—or not to work. We can say that at that time you already had a "social constructionist" approach to technology.

TP   Exactly. But it was not obvious, though, I have to tell you. Things always seem obvious in hindsight, but at that time it was not actually like that. Collins, for instance, was very resistant to this approach to technology back then: he couldn't see the point, he didn't see any relevance in technology at all. He didn't even think he was analyzing technologies: he thought he was analyzing science, not technology. He could not see the point of this program. There were many interesting arguments early on, because Woolgar had also published something quite critical; he thought "Oh, science, to work on science, it's radically revolutionary because no one believes science could be socially constructed. Everybody knows technology is socially constructed, so that point is uninteresting."[4] That was his

---

4. Woolgar 1991, 36:

SST [social studies of technology] has little of the bite and controversy associated with the treatment of scientific knowledge as an analytic object. How many philosophers are going to get upset at the contention that technologies are socially constructed? Not a lot! ... Technological design is not, by and large, an honorific activity championed by a prestigious body of philosophical tradition. The fact that there are no philosophers to upset is disappointing not just because it is fun to upset people, but because ... such opposition is functional for working through the longer term significance of a critical challenge to traditionally held views. Similarly, SST exhibits little of the counter intuition associated with the social construction of science. Whereas, at least at the time of its earliest formulation, the notion that scientific knowledge was socially constructed seemed to contradict commonly held perceptions about science, the suggestion that technology entails social process has little of the same effect.

argument, but I disagreed. We met plenty of engineers who did not want to see technologies as socially constructed either. In hindsight, it seems obvious that all of the concepts we have already discussed in the interview were going to apply, but at that time it was not so clear. I mean, it now seems obvious that you can apply this stuff to technology, but no one had ever published an article in these main science studies journals on pulling apart, opening the black box of a piece of technology.

Even *The Social Construction of Facts and Artifacts*, which came out in 1984, was only published as a discussion paper. I can remember refereeing a paper by MacKenzie[5] in that same journal, perhaps two years after that, and the editor David Edge told me, "If we accept this article, this will be the first article on technology ever accepted. The facts, the artifacts are totally deconstructed inside the black box: no one's done that before," he said. "Should we do it?" So we had a big debate over whether we should publish or not. Nowadays most articles are on technology or medicine or, you know, wider things away from science, but back then it was a real issue: should the journal do it, is it possible to do it? Why would we get involved in this stuff about technology? At the time, somehow, we didn't make that connection strong enough. I think it was implicitly there; but no one had really made the connection yet.

ST At what stage was your career at the time of publishing these two key articles?

TP I was actually in a very strange situation. I had already published many things: my first study was on quantum mechanics,[6] published in 1977. I had worked with Collins on parapsychology, and we had published this important paper on constructing the paranormal[7] in 1979, plus a whole book on that.[8] Then I studied for my PhD on solar neutrinos. While preparing my PhD, I actually had a little center for myself at Bath University: I called it "a center for the social construction of the Sun." I posted this on

For Woolgar, the new object of research, especially when addressed with a "formulaic" approach derived from SSK, represented a diversion from the main goal of achieving a better reflexivity in the field (see Pinch 1993b and Woolgar 1993 for the full debate).
5. MacKenzie and Spinardi 1988a, 1988b.
6. Pinch 1977, 1979a.
7. Collins and Pinch 1979.
8. Collins and Pinch 1982.

my door. I think at this stage of my career, I was a little pissed off. I had a little political goal in this, as well, because I was in a strange situation: this sometimes happens in your career. When I did all the work on the paranormal with Collins, I had been a faculty member at Bath, but when I went to do a PhD I had to change my status. I had to go from being a faculty member to a graduate student. This meant some bizarre things, like I lost some of my rights, like where I could go to eat, things like this ... I think that's why I was making it a little political, saying, "a center for the social construction of the Sun." This was a very difficult stage in my career because there were no jobs in Britain in academia at this point at all. I had this book and the works with Collins, several articles published that were pretty successful in the field, but I was finding, by the time I finished my PhD, there were no jobs at all. I was actually technically unemployed and wondering if I could actually make a career in this, because it is a real shock when you go unemployed.

ST   That is the worst moment in academia: the post-PhD is terrible for everybody.

TP   But, remember, it was even more bizarre for me because I had this odd career with my master's thesis publication, and I had written a book before doing my PhD. I had more publications than most PhD students. And also some credibility, but I still couldn't get a job. And I was really, frankly, pretty desperate at that point. Thatcher was in power, Collins could not do much for me, and there were no jobs. I remember I was at a conference—I was still going to conferences because the field was still alive, particularly in Europe, and there was the first meeting[9] of what became EASST, the European Association for the Study of Science and Technology, which still exists. They had a meeting in a place called Deutschlandsberg, near Graz, in Austria. At that meeting I presented a paper on the social construction of the sun—it was a solar neutrinos work—and there was this Dutchman who I met for the first time, Wiebe Bijker, who presented a paper on the social construction of the bicycle. Now, he'd actually come to that meeting knowing that he wanted to work on the social construction of technology, knowing that there was this movement in the social construction of science, and he had never met these people; he was hoping to meet somebody whom he could get on with, whom he could work with.

9. Sept. 25, 1982.

The blessed thing was that he had some money for this: he had a research fellowship. And I was unemployed. So he gave his paper on the social construction of the bicycle, I gave mine on the social construction of the sun, and we immediately started talking, and he had a background in physics. We hit it off immediately, we became friends almost instantly. He said, "I've got this research fellowship at this place in Twente, in the Netherlands: we could work some more on comparing the social construction of science and the social construction of technology, because I think I have a lot to learn from you." His approach at that stage was indeed rudimentary; it was based on what we call "evolutionary epistemology."[10] He was thinking that technology is kind of like evolution, and he didn't really have any specific ideas about the social construction of science in mind. I said, "I can see how these things could get together. Let's get together and work." And he had the money for it! So he invited me over to be a fellow in the University of Twente. The approach was to compare these case studies. He had different case studies he was working on: bicycle, Bakelite, plastics, weaving machines, lightbulbs,[11] and others I can't remember. We looked through all of these case studies: which one is the most suitable? The bicycle one seemed the most developed and had the most intuitive ideas. He had already developed the notion of relevant social groups around a technology. I suggested, "Okay, what we're going to do is write a joint article showing how the work about the social construction of science could be extended to technology. It could be a joint program that explains how the study of both can benefit each other."

That was important to me because I saw that we could learn things from the people studying technology, but they could equally learn things from us. At about that time there was that article published by Collins about the Empirical Programme of Relativism,[12] and that seemed the natural thing to push. Our article has many aspects: first of all, it is a review of literature.

10. "Evolutionary epistemology" is a label that describes different attempts to account for the innovation of ideas, scientific knowledge or, in this case, technology, through models derived by evolutionary biology and, above all, natural selection. For a critique of evolutionary epistemology, see Renzi and Napolitano 2011.
11. The case studies on bicycles, Bakelite, and fluorescent lightbulbs are included in Bijker 1995, introducing a theory of sociotechnical change based on the social shaping of technology.
12. Collins 1981c. See also Collins 1983a.

It finds any literature that might indicate that this could be a beneficial program. But the intent of the paper is to apply these ideas in the Empirical Programme of Relativism we have discussed (interpretive flexibility, closure mechanism, and Wiebe's idea of relevant social groups) to a particular case study. And we applied it to a bicycle where Wiebe had already done the research, and the bicycle is such an evocative thing. It was just as well that we picked the bicycle, because loads and loads of people have told us that since this work is intuitive, the bicycle examples are beautiful. It is used in teaching all the time. We compared the bicycle with one of my examples from solar physics.

ST   So that is how SCOT took shape. It was a relevant encounter …

TP   It was not called SCOT then, by the way. The name SCOT came later, when the article came out. So when we decided to work together, I could see it instantly. I said "This just needs to be done. We have to write." It was just one of those moments when you know you can write this article instantly because it's basically waiting to be picked off the tree. Why haven't we written this before? It was so obvious! I remember I told him, "We have to do it, it is going to be a killer article." You often don't know if a lot of the stuff you write takes off, if it is going to have any degree of success. But, in this case, I could see it was going to be a killer. Wiebe had not written much in English, so I had to shape a lot of the article, which is why I became the first author of the article. Of course the notion of social construction of technology was his original idea, although it changed as we worked together, and he had the bicycle example, but he didn't quite yet know how to write an academic article, while I had enough experience on this social construction of science to know how to make it work. The name SCOT came out when the article got accepted for *Social Studies of Science*. I was giving a talk up in Edinburgh, in Scotland, and David Edge was there, and he said, "Oh, this is way too long," because it was called something like "A joint program on social construction of facts and artifacts." "We need an acronym. Why not call it something like 'SCOT: Social Construction of Technology.'" That was perfect! Because we were in Scotland, so the acronym SCOT would work beautifully. That led to this famous conference[13] in Twente, and later to that book.

13. The workshop held at the University of Twente in the Netherlands, in July 1984, led to Bijker, Hughes, and Pinch 1987. For a news report, see Pinch 1985a.

By the way, mine was a six-month fellowship, so I was unemployed again. I came back to England and, luckily, I got one of the few jobs in sociology, a temporary job at York University with Michael Mulkay. I was really lucky because I had written a review of Mulkay's book that was just out,[14] *Science and the Sociology of Knowledge*,[15] and he loved it, so he wondered if I was up for a two-year job. That was crucial: a two-year job, giving me enough legitimate employment. I was living in this house in Bath, looking after the three-year-old child of my girlfriend at the time, and I remember picking up the child and telling her "I've got a job!" and dancing around, because it meant that much. By this point we had already done the social construction of technology article, and I knew we had a potentially successful program, but I could still see myself leaving this field because there were no jobs. So getting that position was really a vital moment; two years was really *tremendous* compared to unemployment. I had at least two years, and York is a good university, and working with Mulkay, a very good and well-established sociologist of science, would be a great experience for me.

Meanwhile, Wiebe Bijker had been in Paris. He talked to Michel Callon[16] and realized Callon was also interested in technology; for a while Wiebe flirted with doing actor-network theory, which was very interesting. After Paris, he started to say, "But what about these nonhumans?"[17] He was starting to embrace more of a Callonian approach. I said, "No, we're not

14. Pinch 1982.

15. Mulkay 1979.

16. With John Law and Bruno Latour, Michel Callon is one of the leading proponents of the actor-network theory. See Callon 1986; Callon, Law, and Rip 1986.

17. Sayes 2014, 136:

Within the Actor–Network corpus, the term "nonhuman" functions as an umbrella term that is used to encompass a wide but ultimately limited range of entities. For example, in *We Have Never Been Modern*, Latour (1993: 13) includes "things, objects, [and] beasts" under the heading of "nonhumans"; meanwhile, in *Reassembling the Social*, he includes "microbes, scallops, rocks, and ships" (Latour 2005: 11). More systemically, we can say that the term is used to denote entities as diverse as animals (such as scallops—Callon 1986), natural phenomena (such as reefs—Law 1987), tools and technical artifacts (such as mass spectrometers—... [Latour and Woolgar 1979]), material structures (such as sewerage networks—... [Latour and Hermant 1998]), transportation devices (such as planes—Law and Callon 1992), texts (such as scientific accounts—Callon, Law, and Rip 1986, and economic goods (such as commodities—Callon 1999). What is excluded from the circumference of the term are humans, entities that are entirely symbolic in nature (Latour, 1993: 101), entities that are supernatural (Latour, 1992), and entities that exist at such a scale that they are literally composed of humans and nonhumans (Latour, 1993: 121, 1998).

going down that route." We maintained our SCOT approach, and I think that Wiebe also thought it made sense, although we both were aware of the importance of Michel's approach. But you know: he had been in Paris, speaks very good French, I guess [Wiebe] had been slightly more influenced by him than I was. I had already come across Michel and Bruno's work, but I wasn't quite so enamored with what they were doing. Anyway, that was how the idea took shape. Then we realized that Donald MacKenzie was editing this book with Judy Wajcman called *The Social Shaping of Technology*.[18] There were all these historians of technology like John Staudenmaier,[19] Edward Constant,[20] Thomas Hughes, and Ruth Schwartz Cowan[21] in particular: a whole bunch of historians of technology was suddenly looking for something new. They wanted to tie with sociologists, they were sympathetic toward us. That led to this very successful meeting, which was a breakthrough, and to *The Social Construction of Technological Systems*. And that's how our program took off.

ST   In that early book we already find all the main approaches to the history and sociology of technology. Alongside with SCOT, we already have the systems approach by Thomas Hughes[22] and the actor-network theory, represented by John Law[23] and Michel Callon.[24] What did these approaches have in common, and what were the main differences?

TP   The first thing I should tell you is that in the new anniversary edition of *The Social Construction of Technological Systems*, which has recently come out,[25] Wiebe and I wrote a new introduction which describes our common ground. Unfortunately our third editor, Tom Hughes, was seriously ill, so

18. MacKenzie and Wajcman 1985.
19. John M. Staudenmaier (University of Detroit Mercy) is a member of the Society for the History of Technology and is now editor emeritus of the academic journal *Technology and Culture*, after being editor-in-chief 1995–2010.
20. The historian of technology Edward Constant II was professor of history at Carnegie Mellon University.
21. Ruth Schwartz Cowan was professor in the history department of the State University of New York at Stony Brook until 2002, when she moved to the University of Pennsylvania.
22. Hughes 2012 (1st ed. 1987).
23. Law 1987.
24. Callon 1987.
25. The new edition was published in 2012, on the occasion of the fiftieth anniversary of the MIT Press.

he was not able to participate.[26] But the book has been selected as one of the thirty most influential books ever published by the MIT Press, it is in the MIT Press Museum, so they did a new edition, and we thought this was an occasion for writing a new editorial introduction. We thought about the similarities and differences between these approaches: actor-network theory, systems theory of technology, social construction of technology. The three approaches obviously share a common enemy: a deterministic reading of technology. I mean, people say, "Oh no, there's no technological determinism anymore," but the media is full of technological determinism. Their whole discourse about, say, Internet and smartphones and mobile devices is full of technological determinism. They clearly have a deterministic view, in two senses: first, they think technology is going to impact society in a predetermined way, and second, the actual technology itself has a kind of deterministic logic of technical development that leads it to go in one direction. These two views often come into play together, but all of these three approaches are opposed to that, so they share a common enemy.

They also share the idea that the distinction between the technological and the social, the political and the economic, is something still undefined, up for grabs for people who are developing technology. So you look at Edison, developing his technology of electrical power, and he is not only concerned with working on resistance or lightbulbs, issues that we may now call "engineering technology." He is also concerned with going to Wall Street to raise finances for his companies. So he is juggling with, at the same time, the financial, the technological, the legal aspects. There is a term in the literature for this, developed by John Law early in this field: "heterogeneous engineering."[27] I think the systems approach, actor-network, and social constructions all share some language like that in common. In the systems approach, Hughes talks about the boundaries of systems that are

26. Tomas P. Hughes died in Feb. 2014.
27. "The stability and form of artifacts should be seen as a function of the interaction of heterogeneous elements as these are shaped and assimilated into a network. In this view, then, an explanation of technological form rests on the study of both the conditions and the tactics of system building. Because the tactics depend [...] on the interrelation of a range of disparate elements of varying degrees of malleability, I call such activity heterogeneous engineering and suggest that the product can be seen as a network of juxtaposed components" (Law 1987, 113).

not clear. What counts—the technology, or the economics, or the management—for the system is something that is all up for grabs. Hughes calls it a seamless web.[28]

So the three approaches have those things in common. And I think they are all also interested in "opening the black box of technology," of getting inside a piece of technology, saying whether it is social, economic, political, ideological; just opening the thing up more, showing what goes into it.

All three approaches share that, but from that point on, they start to differ. I think the systems approach has been less influential in STS, but with the recent turn to infrastructure it is making something of a comeback. Well, there are historians who work on systems, so this is maybe my prejudice as a sociologist: historians are still quite influenced by the systems approach. I suppose that the language of the systems approach, whether you like it or not, still has this sort of systems theory ring, from macrosociology or from management theory, somehow at the back of it, and people don't like that; it's something about the word "system." Hughes wrote this marvelous book, *Networks of Power,*[29] where he looks at the development of electricity in Chicago, London, and Berlin and contrasts the different political regimes and ecologies these technologies are developing in. Scholars like Gabrielle Hecht[30] and my colleague Sarah Pritchard[31] here at Cornell have taken the ideas further, showing how the political and the environmental gets worked into technological systems, and in the process have provided a whole new perspective on technology. Actually, also Paul Edwards's work on systems,[32] which is an inspiration for infrastructural studies, and the important work of Geof Bowker and Leigh Star[33] on categorization, can be seem as descending from Hughes's

28. "Heterogeneous professionals—such as engineers, scientists, and managers—and heterogeneous organizations—such as manufacturing firms, utilities, and banks—become interacting entities in systems, or networks. Disciplines, persons, and organizations in systems and networks take on one another's functions as if they are part of a seamless web" (Hughes 1986, 282).
29. Hughes 1993.
30. See in particular Hecht 1998.
31. See in particular Pritchard 2011, 2012.
32. See Edwards 1996, 2010.
33. See Bowker-Star 1999.

work on systems. So I am probably wrong—Hughes's work continues to be enormously influential.

Actor-network theory, which obviously has become huge as an approach, makes the differentiation much clearer because of this radical principle of symmetry. Michel Callon and Bruno, early on, said the human and the nonhuman would be dealt with equivalently.[34] However, I have to say that social construction of technology has evolved; it started off being much more unidirectional: we wanted to explain the social to explain the technical. I think the language these days is much more similar to the mutual construction of technology and society. You will now hear about coconstruction and mutual construction, and also coproduction (as developed by Sheila Jasanoff),[35] so there has been an evolution in the language. I think that everyone who is in the social construction of technology approach would reject this radical symmetry, though. There are issues of material agency to be talked about and worked through, but the idea that somehow you cannot distinguish between humans and nonhumans is something that social constructionists cannot accept. Doing this endeavor, actor-network theory, you start from the assumption that somehow they are ontologically equivalent: it is an assumption that social construction-ists of technologies are not willing to apply. So that is probably one of the key differences.

3.2 Relevant Social Groups, Interpretative Flexibility, Closure

ST Let's go deeper into SCOT, clarifying how it extends some of the key concepts that were tuned up for scientific knowledge, within the EPOR, to technology: interpretive flexibility, relevant social groups, and closure. Of course we have to keep in mind that almost thirty years have passed, and that it is impossible to understand the potentialities of an approach so deeply grounded in empirical research just from its first steps. SCOT is actu-ally not supposed to be "a recipe or a simplistic rule book," or a "formulaic" collection of fixed concepts: it is a methodological approach to producing

34. See Collins and Yearley 1992; Callon and Latour 1992; for a more detailed discussion of the principle of radical symmetry and nonhuman agency, see section 3.5.
35. See Jasanoff 2004.

(and revising) concepts.[36] This must be kept in mind when discussing the main concepts of its original formulation.

TP   We tried to clarify that many times. I can explain these concepts using the bicycle example I usually employ in teaching. So, first of all, *relevant social groups*. Relevant social groups are groups of people to which technology has some sort of shared meaning.[37] SCOT is an interpretative approach to them. They share a meaning about a piece of technology. In the transition from the high-wheel ordinary bikes to the safety bikes,[38] we broadly identified two social groups that shared a meaning: one was the "young men of means and nerve," who ride this sort of bike. By the way, I saw this picture at McDaniel College in Maryland, just few days ago: a beautiful picture of the founder of the college, mounting one of these bikes, and he was exactly this kind of healthy guy who had used this for sporting purposes. These high-wheel bikes were clearly a male bike, used for showing off and for sporting purposes. So, that was one social group, but there was also another social group who wanted to ride bicycles but had problems with

36. "A theoretical approach can never be a foolproof data-mining machine: it will never make the researcher's craft superfluous, nor guard the researcher against errors of using the wrong sources, or of not finding the right sources. The implication is that theoretical frameworks typically are not used without modification and adaptation by the researcher. ... We are indeed happy to report that the scholars whom we know who have tried to apply SCOT have adapted the concepts to fit their specific needs, and most of them finish their study by criticizing some aspect of the original SCOT model" (Bijker and Pinch 2002, 368).

37. "Relevant social groups" must not be mistaken for *groups of users*: nonusers (as well as designers, retailers, etc.) can be relevant social groups for a technology: in this case study, for example, the role is played by young women: "The whole point of introducing the concept of relevant social groups was to get away from such narrow definitions of who and what are relevant in the development of technology. The concept of relevant social group was introduced 'to avoid the pitfall of retrospective distortion.' Our suggestion is that women played a role in the development of the bicycle exactly because they did not use the high ordinary, but wanted to bicycle. Relevant social groups need to be defined more broadly than the standard user groups when one wants to avoid Whiggish history of technology" (Bijker and Pinch 2002, 363).

38. "Safety bikes" are rear-driving bicycles characterized by wheels of similar dimension: as described by Pinch and Bijker (1984), they became popular in the late 1880s among those users concerned with problems of safety of the ordinary bike, characterized by a large front wheel and a small rear wheel.

them, and this was old men and women. In that paper, we classified them together: maybe in hindsight we could have differentiated them, but they had the same sort of problems with the bicycles; they were considered too dangerous for old men and women because of the problem of mounting the high-wheel. Also, they did not want to use them for showing off or for sporting purposes: they just thought they would be good for transport as well. So these are the two relevant social groups that we identified around this bicycle.

We gave an example of interpretive flexibility, which comes from the sociology of science and which means that some aspect of a technological artifact is contested; different meanings are constructed by different groups. So the artifact, in the case of the high-wheeler bicycle, is a contested artifact: for some people it is successful, allowing them to show off and ride, but for other people it is a problematical, dangerous artifact. These are the two meanings written into this artifact. So this is interpretive flexibility: having two different meanings. Over time, in about ten years, the bicycle saw its technological innovation, and there came a safety bicycle: the engineers were responding to the problems of these social groups. They actually responded to both groups, because one response is to build even faster bicycles. They start to build bigger wheels, like the Rudge Ordinary: it has a huge front wheel, the biggest ever. And they're responding to the group of young men of "means and nerve." At the same time, they were starting to work on how to make the bike safer, which led to the safety bicycle.

So: how do you get this closure around this bicycle? We actually thought it'd be neater to show these ideas if we took a clash over bicycle tires: the interpretive flexibility over the bicycle tires. For example, the air tire, which had been developed by Dunlop.[39] This means that there can be interpretive flexibility over some components as well as over the whole artifact. In this case, tires were seen as ugly, sausage-y things, and bike users wouldn't take them seriously. But then, they had been produced to solve one particular problem, that of vibrations. But when they were introduced in bicycle races, air tires made those bikes much faster than the others, which was

39. John Boyd Dunlop (1840–1921) is credited as the inventor of the first practical air tire for "Bicycles, Tricycles, or other Road Cars." His patent was granted in 1888.

relevant for sporting purposes. We talked about this as an example of a closure mechanism which translated the problem of vibrations into one of speed, and thus came to a closure.

ST In your paper you define different typologies of closures, and you call that "closure by redefinition of the problem." You also talk about attempts to come to a closure through rhetorical means.

TP Yes, that is an example of redefinition of a problem. And a certain kind of advertising was an example of rhetorical closure.[40] We tried to identify the mechanisms that led to closure or stability around an artifact. So you have this period of competing bicycles, which lasted only ten years, then a new safety bicycle emerges and becomes a stable artifact. A group of people can ride this bicycle, the bicycle becomes available also for women, and gets called "the freedom bicycle," because there was also an issue of Victorian morals of women riding bicycles that we've thrown in the mix.[41] But those are the key ideas.

ST So this is the basic framework for SCOT: this conceptual framework intends to provide the researcher with a heuristic guide to account for technological change and innovation. But it has relevant theoretical implications as well since it deconstructs any linear and evolutionary idea of technological innovation:[42] where in hindsight you see a linear evolution

40. "An attempt was made to 'close' the 'safety controversy' around the high-wheeler by simply claiming that the artefact was perfectly safe [in an] … advertisement of the 'Facile' Bicycle (sic!). … This claim of 'almost absolute safety' was a rhetorical move, considering the height of the bicycle and the forward position of the rider, which were well known to engineers at the time to present problems of safety" (Pinch and Bijker 1984, 427).

41. Pinch and Bijker 1984, 415–416:

This way of describing the developmental process brings out clearly all kinds of conflicts: conflicting technical requirements by different social groups (for example, the 'speed' requirement and the 'safety' requirement); conflicting solutions to the same problem (for example, the Safety Low Wheelers and the Safety Ordinaries—this type of conflict often results in patent litigation); and moral conflicts (for example, women wearing skirts or trousers on a High Wheeler). Within this scheme, various solutions for these conflicts and problems are possible—not only technological, but also judicial, or even moral (for example, changing attitudes towards women wearing trousers). Following the developmental process in this way, we see growing and diminishing degrees of stabilization of the different artefacts.

42. "In SCOT, the developmental process of a technological artefact is described as an alternation of variation and selection. This results in a 'multi-directional' model, in contrast with the linear models used explicitly in many innovation studies, and

toward a defined goal, SCOT actually shows you a full range of successful and unsuccessful attempts, sometimes conflictive, to answer the problematic nature of the technological object. That's why, earlier in the interview, I was saying that you approach the technological object as a controversial object.

Anyhow, like EPOR, after showing the interpretative flexibility of the technological object and the process of closure, SCOT would require a third phase where the processes described are reconnected to the wider social context. But as for EPOR in those same years, at this stage this is just a little bit more than a promise for future research. I think you will fulfil this part of the program only at a later stage, in your articles and book[43] on the social construction of the Moog synthesizer, working in the field of sound studies. In these more recent works you finally succeed in bringing back the social, showing how the technological innovation in turn shapes social groups, changes their practices, and contributes to redefining their musical cultures.

TP Yes, I think that's developed the most in the synthesizer case study. But Wiebe also develops this a lot in his own book.[44] Anyhow, we don't actually claim we have done it in the case study of the bicycle, do we?

ST You just say that this third phase is needed. Actually, in *The Social Construction of Technological Systems*, where "The Social Construction of Facts and Artifacts" was reprinted three years after it was published in *Social Studies of Science*, you quote Bijker's article in the same book, where the concept of "technological frame" is introduced as a way to address the relationship between the wider milieu and the content of technology.[45]

---

implicitly in much history of technology. Such a multi-directional view is essential to any social constructivist account of technology. Of course, with historical hindsight, it is possible to collapse the multi-directional model onto a simpler linear model; but this misses the thrust of our argument that the 'successful' stages in the development are not the only possible ones" (Pinch and Bijker 1984, 411).

43. Pinch 2003; Pinch and Bijsterveld 2003; Pinch and Trocco 1998, 2002.

44. See Bijker 1995.

45. Bijker 1987, 168–173:

A technological frame is composed of ... the concepts and techniques employed by a community in its problem solving. ... Problem solving should be read as a broad concept, encompassing within it the recognition of what counts as a problem as well as the strategies available for solving the problem and the requirements a solution has to meet. ... The concept of technological frame ... includes such different elements as current theories, goals, problem solving strategies,

TP Yes, and in the case of science, this third, wider-contest stage had not been demonstrated yet. I thought it was demonstrated later by Shapin and Schaffer, with their incredible work on the air pump.[46] Anyhow, Bijker pushes it more for the Bakelite case study in his book, much more than we do in our original bicycle story. It is more developed in Bijker's 1995 book. But as you said, I think that in the case of the synthesizer, one can definitely see the role of wider movements in society. Because we wouldn't have the synthesizer of Robert Moog taking off in the absence of wider movement of the counterculture, which is a wider societal movement, allied with psychedelic drugs. That was the key thing for the development of synthesizers, which was a clearer case study.

ST On the other hand, you showed among other things how the synthesizer, with its powerful sound, contributed to reshaping the roles in a typical band, or how, at a more general level, it worked as a sort of breaching experiment where the ideas of "musician" and of "musical instrument" became shared within a musical culture: the mutual shaping of technology and the social is shown at different levels, both micro and macro.

3.3   Rethinking Users and Stabilization

ST Now that all the main elements of the original SCOT approach have been introduced, we can start to follow some of its subsequent trajectories of refinement and complexification. I would start from the line that has fascinated me the most as a media scholar: the concept of "relevant social group" that brought you to focus on a broader array of social actors

---

and practices of use … and it must be applicable to social groups of non engineers [since] different using practices may bear on the design of the artifacts, even though they are elements of technological frames of non engineers. The concept … is intended to apply to the interaction of various actors … [and structures the] attribution of meaning [of a social group to an artifact] by providing, as it were, a grammar for it. This grammar is used in the interaction of members of that social group, thus resulting in a shared meaning attribution. … A technological frame can be used to explain how the social environment structures an artifact's design. … On the other hand, a technological frame indicates how existing technology structures the social environment.

See also Bijker 1995.

46. Shapin and Schaffer 1985. In their classic, Shapin and Schaffer highlight how the controversy between Boyle and Hobbes over Boyle's experiments with the air pump of the 1660s had political implications that contributed to the shaping of the dispute.

involved in the development of technology and to pay increasing attention to the role of the users in the process. Let's start from the users: in the original version of SCOT, you already underlined their role (as well as the role of nonusers) for the social construction of the technological artifact, but in that first case study on the bicycle, the task of modifying the artifact was firmly in the hands of designers and engineers. In a later case study, on the social construction of the automobile in rural America,[47] you acknowledge a markedly more active role to users. And some years later, in 2003, you and Nelly Oudshoorn dedicated a whole edited book to the role of the users in shaping the technological artifact.[48]

TP Yeah, the relevance of the user was already there in our first case study, but we got criticized exactly for neglecting how users themselves can redesign technologies. The role of the users is there because the relevant social groups (young men of means and nerve versus older men and younger women) were possible users of the bicycles; so we had the category of the users. But this is what we overlooked: we had reintroduced users in terms of interpretive flexibility, but they were not part of a redesign. We hadn't seen that at that point. An article was published by a good friend of mine from the Open University, Hughie Mackay, which was critical of SCOT for not having enough to say about the reappropriation of technology by users.[49] I always try to read these criticisms carefully; they're interesting,

---

47. Kline and Pinch 1996.
48. Oudshoorn and Pinch 2003.
49. Mackay and Gillespie 1992. The article aims at a complexification of the main approaches to the social shaping of technology through a dialogue with the media and cultural studies tradition. Mackay and Gillespie indicate three main weaknesses in SCOT and other contemporary approaches to technical development: an insufficient attention to ideology as functionally and symbolically encoded in the technological artifact; the failure to acknowledge the role of marketing in building and shaping a demand for the artifact and the related relevance of shaping processes after the design phase; and an inadequate attention to the practices of "appropriation" by the users that may involve not only the attribution of meanings, but also practices of direct and unforeseen manipulation and modification of the artifact:

Our discussion of domestic technologies has been concerned with the social forces which are responsible for their development, production and marketing. However, this account is incomplete because it fails to consider the social forces at work on the other side of the technology: the way that technologies come to be actively appropriated by their users. People are not merely malleable subjects who submit to the dictates of a technology: in their consumption they are not the passive dupes suggested by crude theorists of ideology, but active, creative and

you can learn something, and this was a very useful one. "Yeah, what more can we say about users?" Then a historian of technology here at Cornell, Ronald Kline, showed me some photographs: we were over at my house, or at his house—I can't remember—and we were looking at photographs from his album. He showed me a picture of a jacked-up Model T motorcar[50] with a washing machine as part of the rear axle,[51] which we eventually used in our article of Kline-Pinch. I thought, Wow, this is a wonderful example of users reappropriating the technology of the car, giving a new meaning to it, having interpretive flexibility appearing after the design stage. That's why we wrote that article, probably one of the more influential ones on user dynamics within the SCOT program—while, since Ruth Schwartz Cowan's pioneering work,[52] feminist scholars of technology had always been interested in users. And then, Dutch sociologist Nelly Oudshoorn, who had worked a lot on users within gender analysis,[53] invited me to help organize some sessions at a 4S meeting[54] to deal with these users. There was so much interest in it, and we decided to write a book about the whole world of users, giving them more power, you know. And Nelly tied what we were doing back to gender analysis, domestication theory, cultural studies, and innovation studies. Suddenly there was so much interest, other people working on that: people studying technological innovation have been writing for ages about the role of users in innovations. Perhaps not for ages, but at least for three or four more years before we started. For example, there

---

expressive—albeit socially situated—subjects. People may reject technologies, redefine their functional purpose, customize or even invest idiosyncratic symbolic meanings in them. Indeed they may redefine a technology in a way that defies its original, designed and intended purpose. Thus the appropriation of technology is an integral part of its social shaping. (Mackay and Gillespie 1992, 698–699)

50. The relatively inexpensive, mass-marketed Ford Model T was in production from 1908 to 1927.

51. "The farm man's technical competence, rooted in his masculine identity, enabled him to reopen the black box of the car (by reinterpreting its function), jack up its rear wheels, and power all kinds of 'men's' work on the farm and, less frequently, the 'woman's' cream separator, water pump, or washing machine" (Kline and Pinch 1996, 779–780).

52. See Cowan 1983.

53. See Oudshoorn 2003.

54. The Co-Construction of Users and Technologies, held at the Annual Meeting of the Society for Social Studies of Science, San Diego, Oct. 1999.

was this guy from MIT, Eric von Hippel,[55] who published extensively on lead users.

ST This renewed and more detailed focus on the role of the users implied a rethinking of other key concepts of the approach. I am thinking mainly about the concept of closure.

TP Yeah, it had to be complexified. I think that at the beginning we had a too-static, mechanistic notion of closure. I don't like the word "closure mechanism" now, it sounds too mechanistic. Closure is an ongoing process, there are moments of opening and redesigning, reappropriation occurs, interesting things happen at different stages of the technology. So rather than this rigid closure, which perhaps we overemphasized in the earlier days, I'd see it more as a stabilization process: stabilization may be a better term than closure.

ST I think that behind this need for a complexification of the concept of "closure," or for its dismissal in favor of the concept of "stabilization," there are not only theoretical reasons but also historical changes and transformations in the production and marketing of the technological artifacts. At a structural level, post-Fordism establishes a different pace for the innovation process and a different relationship between marketing and production on one side and the variability of the artifact on the other. Both these points were made by Paul Rosen[56] in his discussion paper on the social construction of the mountain bike: he criticized SCOT for

55. Oudshoorn and Pinch 2007, 542:

The detailed research carried out by Eric von Hippel and his students ... has been particularly influential. In one of his first studies von Hippel (1976) showed how users innovate new products in the fast-changing scientific instrument industry. ... After three decades of work, in a variety of product and service industries, von Hippel (2005:2) concludes that users are the "first to develop many and perhaps most new industrial consumer products." ... The users who come up with the innovations, the "lead users," often go on to freely share their innovations so that other users can adopt, comment on, and improve on them. Manufacturers in turn will often commercialize these user-driven innovations. Von Hippel was interested in his early work in how innovating firms can better do their market research to identify lead users. More recently he advocates using lead users as trialists within an iterative process so that technologies and markets are simultaneously constructed in interaction with each other. The model he develops, as is typical for the field of innovation studies, uses quantitative aggregative data. ... This makes the user studies of von Hippel and his colleagues harder to integrate with the other STS approaches ... where "thick description" and more ethnographically inspired methods are the norm.

See also von Hippel 1988.
56. Rosen 1993.

separating the analysis of the broader social context from the other steps of the approach and for making of it a third step. He believes that the analysis should also inform the first two steps.[57] But on the other hand, and from this basis, he underlined that historical change is what brought forth the need to rethink the concept of closure:[58] this means that the differences between Fordism and post-Fordism can also be addressed in terms of the specific kind of closures—or processes of stabilization—they establish. This could be even more true and radical for technologies based on the software/hardware distinction that makes them easier to manipulate. Basically, it seems to me that these technological objects are left as "open" as possible in terms of interpretative flexibility so as to intercept a wider array of different users. Under this point of view, the computer would be paradigmatic.

TP   Yeah, I don't know about that, because we are too close. This is typical of these articles that talk about ideas of technologies like software. In some sense, we are too close to it now to tell. I think that if you'd gone back to the moment of the transition between the high-wheel ordinary and the safety bike, you would have seen a range of different approaches: this is my intuition. For a time, it might look like somehow the bicycles are really

57. "I would argue, consequently, that the distinction made in SCOT, between the first two stages of a technology and the third, is a false distinction. Rather, the social context of a technology, identified through the RSGs [Relevant Social Groups], is pertinent to both the first and second stages of SCOT. In order to understand the full extent to which an artefact can be seen as a 'sociotechnical ensemble' it is necessary to look not just at the internal dynamics of the technology, but to look at the same time beyond this to the wider social world in which they are located" (Rosen 1993, 485).

58. "Since mass production of mountain bikes began, the technology has diversified with the appearance of each new RSG, so that now there is no longer just one 'mountain bike'; rather, there is a different artefact for each RSG, and there appears to be no prospect, need or desire for the stability Pinch & Bijker describe. Transformations in Western society since the late nineteenth century are such that it is now possible, and even necessary, to have a limitless variety of bicycles, and of mountain bikes, available" (Rosen 1993, 505). As a consequence, "the notion of stabilization in SCOT needs more attention. As a start, further research from within the 'new sociology of technology' perspective could be undertaken on the distinction between Fordism and post-Fordism, using artefacts other than bicycles. This would also be a way of assessing how valid it is to introduce concepts such as postmodernity from the field of cultural studies into technology studies" (509).

opening up and that all designs are possible. But you are saying there's some fundamental difference in digital technologies, and I don't know yet. Here's an example: people say, "Okay, we're in an open-source era," right? And then they maybe bring up Wikipedia as an example. But you can go back in time: the solution to the problem of longitude, famously, was a competition in Britain.[59] This is in the eighteenth century. They made a competition to see who could solve the problem of longitude, which means that the whole issue of the best approach toward solving the problem of longitude was opened up. So I'm not willing to say that some fundamentally different mechanisms are at work here. Of course materiality is different. But conceptually, I'm not yet convinced that the process of innovation is wildly different.

The reason I am skeptical about this is that there are always these moments, like in the development of the internet, when everybody is saying that everything is different. This is a well-known part of the hyper-rhetoric of the internet: everything seems different. You get this also from music: everything is different. Well, is it so different? You just get iTunes dominating, rather than record labels. Of course the mode of distribution is different, but the underlying processes remain the same, the record labels still assert their power. iTunes becomes a black-box technology and a different form of distribution of music than LPs and CDs. Over a period of time, you wait for it to crystallize, and you see that things aren't so different. So it's really important not to be taken in by these rhetorics. I'm not saying you're taken in, but I'm always skeptical of claims like that.

Also, I think I've learned enough from my historical case studies and from all the many good historians in my department: I don't want to be guilty of a naïve sort of antihistoricism, the idea that somehow you can use ideas from one period, say medieval times, and just mechanistically put them down in another period and they always catch up the same way. I'm convinced that isn't the most fruitful way of doing it. But I think that, at

59. The Longitude Act (1714) established a public competition with a monetary prize for anyone who could find a practical method to determinate the correct longitude of a ship. While the exact latitude could already be inferred by the altitude of the sun at noon, a technique for the determination of longitude was of key relevance for transoceanic sailing. The prize was never officially awarded, but the Board for Longitude granted John Harrison funds for his research on the marine chronometer.

least, these approaches like SCOT can be useful heuristics as a guide to see the social processes that might be happening there. And they may turn out, the historical work may show, the concepts embedded in a very different way. So I see SCOT more like a heuristic guide. Bijker and I have always said that we don't call this a theory of technology: we call it a heuristic guide toward writing a new sort of history of technology.

ST I understand your point, and I share with you the same skepticism for the hyper-rhetoric of the internet. But actually I was thinking about broader processes, even if of course ICTs [information and communication technologies] play a relevant role there. In particular, I am thinking about the reconfiguration of the mass markets related to new post-Fordist forms of flexible production and to new marketing strategies aiming at a sharper targeting of potential customers. An integral part of these strategies is the diversification of the product, up to all the different forms of customization. All these strategies implement ICTs and technologies in a more and more systematic way to monitor consumers' behavior. So, I was not thinking specifically about the internet as a medium but to these quicker and quicker loops between R&D [research and development], production and distribution, that can happen almost in real time, especially on the internet.

These specific typologies of loops did not exist in the mass market; they were not exploitable in the production process at such a pace. I mean, in the mass-market times of the Model T, you had its users, and these users, as you have shown so well, brought forth unforeseen uses of the technology, deconstructing it. Some of these new uses were supported and encouraged by third parties that started to market kits,[60] some were openly opposed by Ford, like the kit to convert the automobile into a tractor, since Ford wanted

---

60. Kline and Pinch 1996, 786–788:

Several accessory manufacturers took advantage of the car's interpretative flexibility and began to commercialize it. Although firms brought out kits to convert the car into a stationary source of power as early as 1912, advertisements for these kits-and others to convert the car into a tractor-did not appear in large numbers until 1917, during wartime shortages of farm labor and horses. … Firms also introduced more elaborate kits that allowed the car to act as an agricultural tractor during the high wartime demand for farm products in 1917 … [typically] consisting of tractor-like drive wheels, a heavy axle, reduction gears to lower the speed to about three miles an hour, a large radiator, forced-feed lubrication system, and other means to reduce overheating problems. … In terms of SCOT, we would say that … kit manufacturers … [supported] the interpretative flexibility of the car by commercializing what farm men were doing in the field.

to commercialize its own line of tractors.[61] Under some point of view, you are right: today very similar processes are going on. But it's also true that today the producer tries to be already there, ready to observe what is happening to his product and to capture what happens inside the product itself in quicker loops in order to release new or diversified versions. This seems to me to imply new forms of the stabilization process.

TP Quicker loops. It could be, but that seems to me an empirical question. You see, I've written an article,[62] with a former PhD student named Asaf Darr, on selling.[63] We were using Goffman. In the conclusion of this article, we talked about a cross-selling market exchange: the point was exactly mass marketing new technologies that enable you to find out more about things. But what is amazing is that that sector of traditional sales, which sells things in the traditional way, is still growing in the American economy. So my intuition is that, yes, it would be different. You know, the buzzword is "target marketing," right? But people had ways of doing that before. They didn't have these sorts of labels, and maybe they depended more upon informational processes that would pass between salespeople.[64] I am not

---

61. "Although the dealers shared an interpretation of the car with the kit makers and many farmers, they had a subservient contractual relationship with the Ford Company. In this case, one social group used the closure mechanism of contractual power to force another social group to assist it in bringing about the closure it desired.7' One limit to Ford's power, of course, was that Ford dealers could go into another line of business or become dealers for another manufacturer. Dealers who remained with Ford assisted in the restabilization of the automobile on the farm by not selling conversion kits" (Kline and Pinch 1996, 791).

62. Darr and Pinch 2013. For a comparison between selling practices in the mass market and in an "emergent technology market" (3), see also Darr 2006. On hard selling techniques, from street marketing to television direct selling, see Clark and Pinch 1995.

63. "The data which we use in this paper are derived from three different studies conducted by the current authors over different periods: sales of computers and computer accessories in an Israeli chain-store (2007–8), sales of cutting-edge technologies in trade-shows in the US (mid-1990s), and sales of mass-produced consumer items sold by pitchers in open markets in the UK (mid-1980s). Our aim in this paper is to identify those elements of the social organization of sales work which are persistent across the research sites and cultural boundaries, as well as across time" (Darr and Pinch 2013, 1606).

64. In one of their examples about trade shows in microelectronics, Darr and Pinch describe this sort of "target marketing" performed by salesmen:

resisting the notion of changes: these are matters of degree. I am resisting the notion of some huge changes that require a completely different sort of analytical frameworks and tools. This has always been my strategy; I think it is very conservative.

ST To be clear, I do agree with this. In fact we are still discussing these changes, these things that I see as changes, using the same analytical framework. So mine definitely was an empirical question. My point is that it is possible to try to answer in terms of the analytical concept of "stabilization process," especially if we mean it as a dynamic concept. For example, one thing I find very interesting is the proposal of your Cornell colleague, Lee Humphreys,[65] to introduce the concept of "structural flexibility," and her attempt to identify different levels of flexibility and stabilization of an artifact as a way to complexify the idea of "degrees of stabilization," originally introduced by Bijker.[66] It could be used to probe these new marketing and production strategies: the diversification of the object and its stabilization. For example, trying to reread the old example used by Castells to describe what I was calling "quicker loops" among design, production, and distribution: in the early 2000s, Zara was doing in two weeks what Benetton was doing in six months in the 1980s, and this—among the other things—is

---

At the trade show the first engagement with potential buyers happened when they approached the booth. ... The sales engineers strove to identify the clients with the greatest potential to buy. In this cutting-edge industry, where products required adaptation to the specific needs of the clients, this meant to identify the client's application which required as little customization work as possible and was therefore most feasible. To identify the best sales leads the sales engineers took the potential buyers, mostly design and test engineers, through a screening process composed of a series of questions about the application the buyer was working on. Once they had identified a client with a potential to buy, they typically asked him/her to sit down next to a table, usually at the back of the display area, and to go through a product demonstration. (Darr and Pinch 2013, 1615–1616)

65. Humphreys 2005.

66. "[The concept of closure] ... seems to introduce a static element in social-constructivist accounts of technology. Is the process of closure a flip-flop mechanism, digitizing the continuous flow of time? It is primarily to counter this problem that I introduced the concept of degrees of stabilization of an artifact. Following the histories of the various artifacts, growing and diminishing degrees of stabilization can be seen. By using the concept of stabilization in this way, I could argue that the invention of the safety bicycle can be understood not as an isolated event (for example, in 1884) but as an 18-year process (1879–97)" (Bijker 1993, 122).

also thanks to the deployment of a real-time monitoring system of every transaction in the shops.[67]

TP That's a great example. I like that article by Lee Humphreys. She notes that you can have stabilization at different levels as well as pointing to the need to pay more attention to producers as a social group.

ST The thing that seems relevant to me is that the choices of the users— or the customers—become embedded more quickly in the new products. Together with the variability of the product addressed by Rosen in the case study on the mountain bike, it seems relevant for an investigation of the new forms of the stabilization processes; but of course, we are always talking of forms of stabilization, even if the degree of flexibility is different.

TP What you are saying sounds plausible, but I can't say more. I didn't study Benetton, but it really sounds like what I've heard from presentations by von Hippel's group at MIT, exactly in this area of tailored marketing. So, somehow, they want to tie in users, lead users, and then innovate around what users want for very small sectors of the market. But they do it very quickly and responsively. And they were talking about generating products through targeted groups a few years ago; already, it is an actual trend in certain areas. But people tend to forget there's also a lot of mass marketing which is not targeted as well. Car companies still produce pretty much uniform mass-produced car products. Of course they can have adaptations, but still I do not want to be taken in by just a few fashion clubs, even in Milan!

ST Right! But the example of the car is interesting because standardized variability has increased there, too! It seems that this tension between standardization, variability, and quick feedback loops is often balanced through the stabilization of a core, while a belt of other pieces of the technological

---

67. Castells 2001, 74–75:

The secret of [Zara's] success, beside some good-quality design from the great Galician fashion tradition, lies in its computerized networking structure. At the sale points, employees record all transactions in an handheld device programmed with a profiling model. Data are processed, on a daily basis, by the store manager, and sent to the design center in La Coruña, where two hundred designers work on market responses, and redesign their products in real time. These new patterns are transmitted to computerized laser cutting machines in the main plant in Galicia, then the fabric is assembled according to the patterns, mostly in nearby factories. ... In the 1980s, the pioneer of the networking model in the clothing industry, Benetton, had a design/production/ distribution cycle of six months. It was overtaken by Gap when the American firm cut the cycle to two months. Now, Zara does it in two weeks: this is Internet speed.

artifact is kept in a more liquid state. As you can see, for example, in the "disciplined" possibility of customizing software.

TP Yeah, I see this as an empirical question, and, sure, it will be really boring if it all stays the same, but still I don't want to change the heuristic guide. I am not disagreeing with you, but I don't quite yet see the relevance of those changes to the sort of research programs I'm involved with.

ST I don't think this implies changing your approach: on the contrary, I am saying that the approach can be used to read these changes, to address different forms of stabilization through time.

TP Oh! Yeah. In this case I am all for that! Of course, I agree. I'd be interested if someone tried.

3.4 Sellers and Testers (and Sociologists)

ST As we have anticipated when talking about users, another social group you have devoted a lot of attention to in your case studies is sellers. Furthermore, your interest in selling practices is not exclusively connected to your work on SCOT. Where does this attention to sellers originate, and what is the role of sellers in the social construction of technology?

TP I actually became interested in sellers for a project not connected to SCOT. I was teaching at York University in the UK in the 1980s, and there was this strong group of conversation analysts linked to Harvey Sacks.[68] Sacks was a student of Garfinkel, who developed conversation analysis. This group, built around Paul Drew,[69] was very interested in studying conversations. Gail Jefferson,[70] who worked with Sacks, visited a lot—I recall taking Gail to watch her first ever cricket match at Headingley—it was a test match. They did conversation analysis, or CA. Paul Drew and another leading figure, Maxwell Atkinson, switched over to studying institutional talk because it has a simpler structure than ordinary open conversation. If

68. Notwithstanding his early death and his consequently few publications, Harvey Sacks (1935–1975) is regarded as a pioneer in the foundation of conversation analysis. See Sacks, Schegloff, and Jefferson 1974; Sacks 1972, 1995.

69. See Drew and Heritage 1992 and the recent reissue of the four volumes of *Conversation Analysis* (Drew and Heritage 2013).

70. During the 1970s, Gail Jefferson (1938–2008) launched conversation analysis with Emanuel Schegloff and Harvey Sack, and introduced the system of notation for transcribing talks still used in the field.

you just looked at institutional talk, say, at law courts[71] or news media,[72] there were more constraints, so they could say more about the turn-taking properties, the linguistic properties, the rhetorical properties of talk.

I was quite influenced by that group, and I wanted to learn, because they were using videos, they did this interesting stuff, and I was keen to learn something from them. So I teamed up with a grad student of Paul Drew's, Colin Clark,[73] and we studied a group of market sellers who were using speech to pitch their goods at the market, a very traditional way of selling. We did a big study on these groups around markets, and this participant observation and video study led to a book that came out many years later, in 1995, called *The Hard Sell*.[74] The study involved two video cameras, one on the seller, and one on the audience. We studied different markets in England, in the North and the South, and we've got some data from America, some also from France and the Netherlands.

At that time, of all those people studying institutional talk, I was most influenced by Maxwell Atkinson, who published *Our Masters' Voices*,[75] which is a very nice but underknown study of political rhetoric. He looked at people like Margaret Thatcher and Arthur Scargill, the leader of the National Union of Mineworkers at the time, and focused on the applause: the trick he did was to look at the speech and gestures and see how the audience applauded in response. So this made it easier to study because the response was simplified: applause or no applause, lengthy applause or small applause. I thought that was a really clever approach because you could study bits of particular rhetoric and gesturing and listen to the applause. It is worth saying this because, though it was just such a simplified thing, it was really important: he found two major rhetorical devices that generated bursts of applause—one was the *list of three*, at which Margaret Thatcher was really superb. She said, "Soviet Marxism is ideologically, politically, and morally bankrupt," and she did this hand gesture at each point, a list of three. It has been known since Aristotle, but that was a sort of Aristotle for postmodern technology. The other is the *contrast device*, he calls it "the

71. Atkinson and Drew 1979.
72. Clayman and Heritage 2002. See also Drew and Heritage 2013.
73. Pinch and Clark 1986. See also Clark and Pinch 1992, 1994; Clark, Drew, and Pinch 1994, 2003.
74. Clark and Pinch 1995.
75. Atkinson 1984.

contrast": two things in tension. The political speeches you tend to remember often involve contrast: Kennedy's "Ask not what your country can do for you, but what you can do for your country" is a classic example of contrast.

So we were studying these market sellers, and instead of producing applause they were producing sales. The idea was very simple: let's study these very traditional forms of selling and see which sorts of rhetoric drove the sales. But it got more complicated: it wasn't just the three-part list and the contrast devices. Working out what was going on turned out to be rather more complicated. But we persevered, eventually working out the sorts of techniques they were using; how they constituted obligations in selling; how they persuaded people to part with their money; and so on. So I didn't work on selling: I worked on the rhetorics of selling. At that point, I knew the sellers were understudied because I had read the literature on selling, and they were hardly ever being studied. But I was going to put it together with the work on technology: that was the strange thing.

At that point of time, studying the rhetoric of selling was sort of a separate project, but I knew sellers were important. So when I came to study the Moog and write my history of the Moog, I knew intuitively that it was important to look at the sellers, who have always been ignored. When I realized at some point that music synthesizers get into retail stores, I wondered, how did they get there? I knew I had to find the people who did this. I tracked down this charismatic salesman, David Van Koevering,[76] an ex-evangelist preacher. He was the key guy, the one who got the synthesizer

76. "It was 1970 and Moog's small Trumansburg factory [...] in desperate financial straits and under much pressure, had brought out the Minimoog—their first portable keyboard synthesizer. It was set to become a classic but they, as yet, had no clue who would buy this synthesizer or indeed whether it would sell at all. Even Moog himself couldn't fathom who the customers might be. Minimoog production was started in batches, just enough to meet current orders. Then something happened. One salesman started to buy more and more Minimoogs; furthermore he was selling them in retail music stores. That salesman was David Van Koevering" (Pinch and Trocco 2002, 237). "In 2 years, Van Koevering went from a demonstrator of novelty musical instruments to vice-president of the Moog synthesizer company. It was Van Koevering who first saw the new instrument's potential in rock music and who took it on the road in a Cadillac to recruit a network of dealers, first in the United States and later throughout Europe and around the world" (Pinch 2003, 252–253).

from the factory into the market, building up the whole retail network. For me it was a really important thing to focus on. I started to see the seller as the key intermediary between users and producers and started to think about how these intermediaries work.

The great thing about sellers is that they are mobile; they are moving backwards and forwards. They are moving around the country, in the case of America, establishing sales networks. They are very mobile people. And they are also moving between the manufacturers and the consumers of technology. This mobility is what gives them their power. They are also another understudied group, so I wrote this article, on how the Minimoog was sold to rock 'n' roll,[77] which focuses in particular on the role of this particular key seller. He is an innovator as well because he comes up with this great idea. He realizes, and he certainly had the position to do it, what the users lacked: they cannot understand how to set up this complicated instrument. Think of a violin being sold for the very first time, in a music store, without a canon of violin music. You've just got a violin. You go to a music store, no one has ever heard a violin before, and you say "I want to sell this in your store." And the guy says, "What can you do with it?" And you play a tune, but then he tries it, and he can just squeak it out. "It is really hard to play. I can't sell this thing." The synthesizer in its first keyboard form—which had, like, forty-four knobs and switches—is impossible to sell. The guys in the store say they can't sell this, it's too complicated, too difficult. You know, there is nothing except the sound it produces, there is no musical canon, there are very few artists using it.

So what does Van Koevering do? He devises a brilliant way of selling it and, also, simplifies it so that users can use it. This is a simple material practice: he carries this tape with him and finds users in the clubs. He goes to a club, finds the people who are playing Hammonds or other sorts of keyboards. "Try the Moog keyboard: the Minimoog!" He then lends them one and teaches them how to set up the sounds by using these little bits of tape: "You know, that's a great sound, we're going to put all knobs on red so you can make that sound again. Got another sound? Okay, we're going to put a little bit of tape: that's the green sound. Then anything else you find in between, we put a label on that." So he devised the whole idea, which is a really crucial one, that these sounds can be emulated and repeated.

77. Pinch 2003.

Eventually this becomes a special sound chart of all the things people can draw on. As he was the one who came up with the idea, this was user innovation from a salesman!

The idea gets back to the company, and they say, "Okay, we'll produce these cardboard charts and a better manual on how to set up these sounds." Then, at the same time, he has devised a way of selling instruments in retail music stores. And what he does is convince these users that they've *got* to have these instruments. He sets up the finance package for them, which is not insignificant: he works on the one person who can buy the instrument (usually the girlfriend, or the mother, or the family, because a guy in a band has no money)! He takes him to the music store. This is the brand new thing he does, this new form of demonstration. Rather than signing the deal with the guy, he takes the guy to the store and shows him to the store owner: "Here's your market." A nice twist on demonstration. "Here's the guy, he is ready to sign. I am going to sign a deal with you where I'm going to supply you with these synthesizers. Here's your first customer, and there'll be other customers coming." Basically, this is what I wrote about. It seemed to me that sellers were kind of a missing link in how technologies get out there into the world, a key intermediary. I don't know if anybody has written about sellers since, but I think they should.

ST   The book that comes to my mind is the one on the VCR published by the MIT Press and written by Joshua Greenberg.[78] It describes the birth of the market of movies on tape while the VCR was being marketed as a technology to record TV shows—a device that could free users from television schedules. He focuses on the intermediaries between manufacturers and consumers, including sellers, and he extensively quotes *Analog Days*. I think *Analog Days* is not only a great book on sound studies but also the one where you have shown better, at least so far, the full potential of SCOT. You have this very articulated map of relevant social actors, you have these continuous stabilization processes described at many different levels, and you have this coshaping relationship between the instrument and the users—the bands—that allows you to address the relationship of the processes you describe with the wider social context.[79] To use an example that we have already mentioned, you show how Van Koevering could foresee a new role for the keyboard player in a band: he could finally

78. Greenberg 2010.
79. See also Pinch 2008c.

compete with the guitarist. The Moog was reshaping their roles and music as well.

TP Yeah, Van Koevering hears the Moog sound. The sound is really important here as well. Because he hears the Moog, not the Minimoog, which had not come yet: he hears the modular Moog. The amazing thing when I interviewed this guy, he was talking about events thirty years earlier, but it still brought tears to his eyes. He is describing these things, and he becomes all emotional: "I heard that sound. And the sound was powerful, and I could see what this sound could be in the hands of these young guys. It could be everywhere. The sound was so powerful." He is this kind of person: he was not really a "visionary," he hears it really, and he tries to translate that into a way, a material way of doing it and he does, he succeeds. So I kind of think the sales people, they've always been overlooked. They are seen as just having the gift of gab. Of just being orators. They are kind of sneered at because sales peoples' job is to sell stuff. But their job is actually also a creative one. I think if you look at the creative activity, you get a better understanding of what these people do and why is it so important. That's what I am trying to do.

ST You have shown that when you open the black box of technology, you can also find their hands there, their fingerprints, together with the ones of designers, engineers, and users. Who else's hands did you find there?

TP Repair people. Repair people are interesting because, in the case of the synthesizers, they are really wonderful examples of this. This is a good example: the first commercial digital synthesizer comes out in 1983. It is called the Yamaha DX-7 synthesizer,[80] and it's very complicated. It's polyphonic, it makes a beautiful sound, but it's really hard to set up sounds on. So what they find is that when the synthesizers are coming into the Yamaha factory to the repair people, what they all have on them are the factory presets: people have not tried to set up new sounds because it's too hard. So the repair people learn from this and tell Yamaha: "This synthesizer is too complicated, people cannot make their own sounds on it any longer." People used to make their own sounds. So then Yamaha realizes this, and what they do is, they realize they are going to have to stimulate a special outsourcing. They are saying, "We need a cottage industry that would do programs for this instrument." So they start to hold competitions, organize

80. See Pinch and Trocco 2002, p. 317.

competitions for people to make sound patches for the DX-7, and suddenly this industry of people making these patches on soundcards, which can go into this synthesizer, is established. People learn to program it. But not themselves, they're buying the programs from independent people. It was repair persons who first discovered this. This is how the story goes. So I think repair people are of key relevance, again because they're intermediaries between the factory and the users. So I am looking now at repair people, salespeople, marketing people: these intermediaries. There may be others, but these are the ones I have looked at.

ST Yeah, this would be also of key relevance for media studies, where the main analytical attention is usually focused on users, and these intermediaries are definitely understudied. Together with sellers and repairers, at different points in your work, you have also stressed the relevance of another social actor: testers, people who test technology. This is another understudied field in media studies. What can we learn about technology by looking at testing and testers?

TP I am trying to remember exactly how I've gone into testing since I have started looking at technology. I think it was when I was at York: we were doing that study of clinical budgeting we have already discussed,[81] the health economics work, and we studied demonstrations and tests of that clinical budgeting system. At the same time I was reading about the *Challenger*, the Shuttle *Challenger* explosion.[82] I realized then, reading the Presidential Commission report into the accident, that there were also

81. See section 1.5.
82. The *Challenger* Space Shuttle was operational from Apr. 4, 1983 to Jan. 28, 1986, when it exploded seventy-three seconds after its launch. All seven members of the crew died. The commission appointed to investigate the incident, headed by William Rogers, imputed its causes to the loss of resilience at low temperatures of the O-ring seals used in the solid rocket boosters. "Worse, it emerged [from the work of the commission] that engineers from the company responsible for building the Solid Rocket Boosters, Morton Thiokol, had given a warning. At an impromptu midnight teleconference the night before the launch, they argued that the O-rings would not work in the bitter cold of that Florida morning. The engineers, it transpired, had been overruled by their own managers. In turn the managers felt threatened by a NASA management who expected its contractor to maintain the launch schedule" (Collins and Pinch 1998, 30–31). See Vaughan 1996, the main reference of the case study discussed in Collins and Pinch's *The Golem at Large* (1996).

some tests, in this case of the equipment that's gone wrong: the pieces they recovered from the accident.[83] And that played a role. Then there was MacKenzie. Donald MacKenzie, at this time, in a volume I coedited called *The Uses of Experiment*,[84] had written an article about nuclear missile testing[85] using ideas from Collins:[86] focusing on similarity and difference judgments that are made in tests, which is an idea behind this repeatability

83. For Collins and Pinch (1998), the results of these tests could be considered crucial clues of the impending behavior of the O-rings only in hindsight, since they required a judgment of similarity between the previous situations and the present one that could not rely on any objective criteria:

On the night before the fateful launch of Challenger, a teleconference was held between engineers at Morton Thiokol, the manufacturers of the shuttle's solid rocket fuel boosters, and the National Aeronautics and Space Administration's Marshall space flight center. At issue were some results put before the teleconference by Morton Thiokol engineer Roger Boisjoly. These results indicated that there would be a risk of O-ring impairment if the shuttle were launched at the low temperatures expected the next morning at Cape Kennedy. Boisjoly's argument in its essentials was that O-ring blow-by observed from the boosters recovered from previous missions was correlated with temperature. Boisjoly's claim ... was that there was enough similarity between previous results at low temperature ... and actual use ... to project that low temperature could lead to O-ring blow-by. ... However, the validity of this conclusion was questioned during the course of the conference when, among other things, it was pointed out that O-ring blow-by had also been observed on flight SRM-22, which had been launched at a high temperature. (Pinch 1993a, 34)

Consequently, the authors advocate a redistribution of the political responsibilities of the incident:

The space shuttle had been presented to the public as a safe means of transport, its security symbolized by the fact that a school teacher was to ride in it. The responsibility for the shock and horror lies with those who, in the modern political blaming jargon, provided the 'false prospectus.' Those people, however, are uncomfortably close to the top of the political power structure ... so it is perhaps not surprising that a large number of people are happy to maintain the fiction that the Challenger was about as perfect a means of transport as an airplane and that something went wrong was the fault of someone lower down in the hierarchy. (Collins and Pinch 2007, 254–255)

84. Gooding, Pinch, and Schaffer 1989.

85. MacKenzie 1989, 430:

Testing is crucial. But the knowledge that testing generates, like that generated by experiment, is always potentially problematic. The auxiliary hypotheses involved in testing, the competence of the testers, the judgments of similarity involved in making inferences from testing to use, the removal of modalities involved in the construction of test-based facts—in principle all of these are always open to challenge. Matters of social interest. Expectations and conventions—issues of credibility—enter into what challenges are actually made and what their outcome is, and thus enter into the construction of knowledge from testing. ... The analogy with experiment is strong.

86. See Collins 1975, 1985.

idea of Collins. It's problematic when two things are similar. It goes back to Wittgenstein: what do you think is different? Similarity works and difference works as interpretive work, so that is Wittgenstein's point. So when two experiments are the same it depends upon you, as an analyst, finding these things that are similar between them—or you could say that they're different. And this is a judgment that always relies on assumptions by analysts and by participants as well.[87]

So, the testing idea came from thinking about that in terms of what's being actually tested, in the space shuttle, but also in sonic technologies: there are always similarity assumptions. I was working on this, and I got the idea of projection, which I think was a nice way of putting it: that somehow what's going on is a process of projection from the test situation to the use of the actual technology. When the roadie checks, or when you, before the interview started, tested this microphone—"Check! One, two, three"—you assume and project that in half an hour's time it is working the same way. So there's always some sort of projection of space or time involved, and it depends on the shared assumptions within a social situation, or a community, as well. It seemed to me that was worth writing about. Testing is the equivalent of the experiment, for the technology. So in that article I wrote about that, and you can think of it like "This is one way how engineers produce knowledge."

Then it got even more interesting. I have got to tell you, at that time I was working on testing—I think it actually could have been around 1990— I'd written a paper a little earlier that remained unpublished. I was in Berlin in 1990, working with Bernward Joerges who had invited me to be part of his research group at the WZB[88] for the summer. The wall had just come

---

87. Pinch 1993a, 31:

In technological testing, the same point [described by Collins's "experimenter's regress"] arises in principle. It is possible to refuse to accept test results by questioning the similarity relationship needed to make the projection from test to 'real-world' use. Unlikely as it is to be significant, it is nevertheless the case that there is a difference between the roadie holding the microphone and Mick Jagger holding the microphone. Of course, whether this difference is considered significant depends on a lot of other assumptions. Its relevance could be made more plausible by pointing out, for instance, that the physical vibrations produced by 20,000 rock and rollers gyrating at the sight of Mick Jagger holding the microphone may make it fail on that night. In other words, the test results can be challenged by pointing to the potentially deleterious effects that such vibrations may produce.

88. Wissenschaftszentrum Berlin für Sozialforschung, the Berlin Social Science Research Center.

down. Western cigarettes started to be imported, and they had a big advertising campaign aimed at East Germans. I guess British Imperial Tobacco
is the company, the backing company behind it, and the brand was West
cigarettes: "Test the West." They had all of these crazy advertisements. You
can find them online. I started to collect these: the "Test the West" cigarette
ads. They have all these people in bizarre situations, like this Elvis character, somehow smoking a cigarette, and it's a test of the West. So I started to
analyze this in the context of cigarette advertisements, and then I got the
idea—"Hey, testing can be even wider than technology"—because I started
to realize that the language of testing was used increasingly in the media
and politics. You find that George Bush, after the first invasion of Iraq, said
famously, "This is a test of the new world order"; so the idea of a "testing
discourse," or "testing rhetorics" is used in politics as well. We are in a kind
of testing society, where more and more about individuals are tested;[89] tests
of your blood pressure, your blood sugar level, SAT tests in America,[90] and
IQ testing, which was the start of it. More and more tests.

So in this world of tests, I started to think: Why is it that even in politics
people are starting to frame things in terms of, "This is a crucial test?" But
we are doing it everywhere, every day: "This is a test of our relationship."
So I got interested in this wider rhetoric of testing as well. In one of my

89. Pinch (1993a) distinguishes four different kinds of tests: prospective ("when a
new technology is tested before being introduced"), current (a test that "occurs once
a technology is up and running ... to assess performance, to make improvements,
to compare it with rivals, for legal and safety reasons ... or to ascertain any special operational difficulties"), and retrospective (occurring "after a major accident or
malfunction has occurred in a piece of technology"). A fourth kind of test aims to
project the future behavior not of the technology itself, but of the user ("any technology that requires the user to act in new sorts of ways ... will involve some in vivo
testing. This is because the manufacturers cannot be sure that the users will be able
to do what is required of them") (36). In this case, presumptions about the user—
including social and political presumptions—would be "black-boxed" in the
machine, with relevant political implications: "The danger is that, unless such
relationships can be contested, one particular version of the user will increasingly
be incorporated within technoscience systems" (Pinch 1993a, 37). It would be one
of the main duties of a "sociology of testing" to contribute to the opening of this
specific kind of black box.
90. The Scholastic Aptitude Test (SAT) is a standard test used to assess students' academic readiness for college and is widely used in the United States as an admissions
test.

papers I actually tried to frame it in terms of Luhmann's ideas, because the sociologist Niklas Luhmann, in one of his essays in *Trust and Power*,[91] had written about how we move from individual trust to systemic trust. So society's gone to this stage where it's moved to systemic trust. You used to trust individuals, now you trust the bank, the system, or the institution. But system trust is breaking down, with institutions such as banks failing. I had this idea that, in that sort of situation, testing was one way of somehow dealing with a lack of political trust in institutions. That now, individuals can test continuously these institutions, like the test of the new world order. They construct this uncertainty about institutions breaking down as a form of testing. Now that's my idea, and I'd like to further develop it: it goes beyond technology.

ST   What about sociologists? Do you think that today they can be considered a relevant social group for technological development? Are they playing a role in technological development and innovation?

TP   Well, they are. Just having a sophisticated take on technology and technological innovation, the sociologist has something to offer, using these sorts of theories, and SCOT and actor-network as well: all of these more sophisticated takes on technology. If you want to understand how technology is going to develop, what the problems have been in the past, what they are going to be in the future, whether there would be a return to invest in these technologies or not, this approach would be of some help at some level. It gives you a framework for understanding technological development and innovation. Sociologists are also hired as ethnographers and anthropologists, in certain technological fields, to try and understand what's happening. Google is often doing this. Intel has got quite a few sociologists working on projects like which way, or direction, the internet is going.

I have to say that there's always a danger, because they can get somehow captured by the organization and then the findings that they produce are not so interesting. Provided they have a way of maintaining what is called an *independent analytical integrity* using their knowledge based in the academy to understand what's going on, they can be very useful to these companies as well. But when they get captured, they're less useful, actually, because they're taken in by the kind of bureaucracy of the company and

91. Luhmann 1979.

they start to express their findings in those terms. The very social process they are studying somehow takes them over. I think that Lucy Suchman, who worked for years in Xerox PARC, had something to say about these issues.[92]

Obviously, sociologists also play a role in education, for science and engineering students,[93] for example. Sociologists can intervene specifically in the science curricula, and the technological curricula, and even possibly in the medical curricula, trying to train people in a slightly different way, with more attention to the processes of these technological endeavors or scientific and medical endeavors. These are the roles they are playing.

3.5 Materiality and the Nonhumans

ST After this overview on the role played by different typologies of relevant social actors in the social construction of technology, we are getting closer to your present work, and to SCOT's most recent developments. One of the main new themes seems to be the theme of materiality.

TP Well, I don't think this is something new for SCOT. Maybe it is still quite new for sociology, although in some senses it shouldn't be, because it's hard to think how you could have a sociology of anything without considering materiality. So there is this interesting question of whether classic sociology has in fact considered materiality and it has just somehow vanished from today's sociology. That is one project one could imagine doing, and that's in fact the project I did with Goffman's early work[94] by looking back at *Presentation of Self in Everyday Life*.[95] I realized, in a particular section on regions,[96] that Goffman was well aware of the material constraints on interaction. He was interested in how regions are materially formed. This

---

92. See Suchman 2007.
93. Pinch 2008b.
94. Pinch 2010a.
95. Goffman 1959.
96. "A region may be defined as any place that is bounded to some degree by barriers to perception. Regions vary, of course, in the degree to which they are bounded and according to the media of communication in which the barriers to perception occur. Thus thick glass panels, such as are found in broadcasting control rooms, can isolate a region aurally but not visually, while an office bounded by beaver-board partitions is closed off in the opposite way" (Goffman 1959, 106).

is the basis of the front stage and backstage distinction. He was very interested in radio: radio studios had glass walls that allowed vision through, visual information through, but not sound. He contrasts it with American offices made of beaverboard: this thin board of the time that cut off vision but allowed sounds to travel. So he was very aware of materiality playing into performances or the "interaction order," although he didn't stress it as a theme.

In my material reading of Goffman I pointed out that Goffman was actually a sociologist of technology, because he was interested in this door between the front stage and the backstage of the Shetland Hotel.[97] It is the most contested space in his ethnography of the Shetland Islands, where he develops the front stage and backstage distinction. The waitresses want the door open because they want to go through without having to be impeded because the door is closed. They bring the dishes, so it is not easy to open the door. There is a whole material arrangement here: they carry a tray of food, they can use only one hand, and they do not want to push this door open, since with one hand this is awkward. But the manager of the restaurant wants the door to the kitchen closed because he doesn't want people in the front stage, in the middle-class parlor, looking into the kitchen, where the working class ambiance is. So you have a class issue and a gender issue, because you have the waitresses and the middle class male manager. Goffman beautifully points out that there is a technological solution to this with the kick-door. The kick-door has got a glass panel, that allows you to see through, and you can kick it with your foot: the door would open, and then it would slowly close. That's why Goffman is a sociologist of technology, I argue.

97. Goffman 1959, 118–119:

Given ... the various ways in which activity in the kitchen contradicted the impression fostered in the guests' region of the hotel, one can appreciate why the doors leading from the kitchen to the other parts of the hotel were a constant sore spot in the organization of work. The maids wanted to keep the doors open to make it easier to carry food trays back and forth, to gather information about whether guests were ready or not for the service which was to be performed for them, and to retain as much contact as possible with the persons they had come to work to learn about. ... The managers, on the other hand, wanted to keep the door closed so that the middle-class role imputed to them by the guests would not be discredited by a disclosure of their kitchen habits. Hardly a day passed when these doors were not angrily banged shut and angrily pushed open. A kick-door of the kind modern restaurants use would have provided a partial solution for this staging problem. A small glass window in the doors that could act as a peephole ... would also have been useful.

Obviously all science and technology studies approaches, in opening the black box of technology, are dealing with materiality: it just seemed so obvious that, early on, people didn't use the word "materiality." I guess one of the turns has been Latour's looking at infrastructures:[98] there, the materiality also plays out as being very important. That's one of the places where people start, in sociology, to notice materiality and issues of infrastructure. So for me it's not a new thing, for SCOT or for my research; it is just a new way of emphasizing something that is missing in this continuing debate with traditional sociology. Early on I thought we stressed that because that seemed to be one of the things that was missing. Now it seems there's even something called "the material turn": everyone suddenly turned to those things that were in the air and suddenly went into materiality. You know, there's this anthropologist called Appadurai who has written about the material life of objects:[99] you find different fields that are interested now in materiality. And also there's a huge audience, I've found, in organizational sociology, in business schools, which suddenly begin inviting me to give speeches on materiality. It's odd because, for me, it is sort like a return to an old theme. I think in your field it could be different; in internet research and new media studies, it does have some relevance in thinking about the materiality of interaction,[100] but in general, it's just part of the SCOT message. For me it's a reemphasizing of something that has always been there, for a new audience.

ST   This is interesting: I mean, fields of research becoming aware of the relevance of aspects of their object that have been already explored in contiguous fields at different times, starting a sort of nonsynchronic dialogue that sometimes forces the scholars in those fields to reemphasize their earlier work. And, I have to say, that actually is exactly what I want to do with you, since I'd like to ask you to go back to your work on materiality and infrastructure, to clarify the specificity of your approach to the topic.

TP   Ah, ah! Well, the infrastructure example is very well developed in the article I wrote on putting the nonhumans to rights,[101] which was about

98. See Latour 1988b, 1992; see also Latour 1996 on the Aramis project in Paris (a personal transportation system, finally put aside in 1987) approached through the lens of the actor-network theory.
99. Appadurai 1986.
100. On online interaction, see Pinch 2010b.
101. Pinch 2010a.

the infrastructures in my own political municipality of Forest Home, where I was for a time the president of an organization, called the Forest Home Improvement Association, which is essentially a local organization of house owners that addresses issues around infrastructure such as roads, sound, things in the environment like animals, or nuisances or benefits in the environment. So I've used my own position in this political debate over measures introduced to calm traffic in our little village of Forest Home. I was there when the political negotiations took place. In fact I took part in the political negotiations over exactly what types of material we would use in issues like slowing down cars. Whether we wanted cyclists to ride in the main road, or to the side of the road, became dependent upon a certain sort of material used on the side of the roads that would give the cyclists so-called buzz vibrations, forcing them into the middle. In the paper I published in the *Cambridge Journal of Economics*, I use it as an opportunity to examine more closely the view on nonhumans in actor-network theory.

Although actor-network theory, Latour and Callon, agree with the social construction of technology on many, many things, I think their approach is a little too radical in treating the humans and the nonhumans as being equivalent:[102] a full symmetry. I actually do not believe that, in their studies, they do this, so it's a silly methodological principle, because they don't follow it; because, as I pointed out in the article, there are simply too many

102. "Our general symmetry principle is ... not to alternate between natural realism and social realism but to obtain nature and society as twin results of another activity, one that is more interesting for us. We call it network building, or collective things, or quasi-objects, or trials of force ... and others call it skill, forms of life, material practice" (Callon and Latour 1992, 348). The principle of radical symmetry was originally introduced as an alternative methodological approach to the role played by nature in the production of scientific knowledge. As discussed in section 1.4, methodological relativism brackets as unaccountable the role played by nature in the closure of a controversy, prescribing to the researcher a focus on the social aspects of the process. In contrast, the principle of radical symmetry circumvents the problem of nature's agency, holding the distinction between "social" and "natural" as an effect of the process of knowledge production itself. As such, the distinction should not represent an a priori methodological guide for the analysts, who should instead attempt to treat symmetrically social and natural, or human and nonhuman, actants under the point of view of agency. For a discussion of the affinities and divergences between actor-network theory and Algirdas Julien Greimas's semiotics, see Akrich 1992a.

nonhumans in the world to follow them all. So you have to have some metarule, or rule of thumb, a heuristic guide, as to which of these non-humans to focus upon.[103] For example, it is taken for granted that when we talk about the bicycle it has the force of gravity acting on it. Everyone knows that bicycles would fall over if you dropped them. The reason why the wheels stay on the ground and they are not floating in the air is because gravity is acting there. But you don't need to go into the laws of gravity: you can keep that black-boxed. What I am trying to say is that there are a lot of nonhumans "acting," and there is a lot of taken-for-granted knowledge about the material world, but you don't need to unpack it all, because only certain things are at stake. What things matter depends on the particular situation, and on how the actors view the thing. This is also the answer that I gave to Peter-Paul Verbeek's allegation that SCOT cannot cope with the material world.[104] He claimed that SCOT doesn't deal with material effects but only with meanings, and he was using the example of the microwave oven,[105] taken from a book by Cynthia Cockburn and Susan

103. Pinch 2010a, 82:

If you think of the non-human effects in even describing how a bicycle works there are clearly many: the frame must be rigid enough to support riders; the handle bars must turn freely such that riders can not only steer but also react to bumps on the road; the chain must be tight enough to enable a hill to impact on riders such that they feel the need to pump their legs harder; the tyres must be inflated and of the correct material to lessen impacts of bumps in the road and so on. Add in all the laws of physics and balance so that upright riding can be maintained under different conditions and you have vast numbers of non-human effects. In writing about the bicycle Bijker and I were, of course, quite aware of the many ways that bicycles, roads, and so on, impact humans. We saw no need to lay stress on such effects, but neither does our form of analysis deny such effects. Rather our analysis is selective in the aspects of the non-human world it chooses to focus upon.

See also Pinch 2009.

104. "Limitations were soon found in the SCOT approach as it became clear that the technologies themselves also played an active role in 'social' interaction processes. The example of the microwave oven … is a beautiful illustration: the factors that determine whether human beings take their meals together include not just human beings but also the microwave itself. Reducing technology to social interactions therefore fails to do justice to the active role played by technologies themselves" (Verbeek 2005, 102).

105. Verbeek 2005, 5–6:

The microwave has [a] gender-stereotyping effect not so much because of its materiality but primarily thanks to its meaning. The original microwave was a symbol for technological sophistication because of its complexity, while the newer version symbolizes technological illiteracy by virtue of the low-tech places where it is sold and its simplified operation. When the microwave

Ormrod.[106] Susan Ormrod was a master's student of mine at York years ago. I really love the book she did with Cynthia on the microwave because it has one of the best chapters on how technology gets gendered in the process of sales and consumption.

So, in that article I said that the humans are dealing with nonhumans particularly in these political forms, such as in the Forest Home Improvement Association. Political negotiations over infrastructures: this is a reasonable way of finding which nonhumans are at issue. I used that article for making that sort of point, and also I realized in writing that article that Latour's famous example of the speed bump, where he talks about morality being inscribed in materiality,[107] a very famous piece, raises similar questions. He uses other examples, like door openers and closers and seat belts. Then I suddenly had this insight that this issue is no more than the issue of your bicycle being forced around the corner by a piece of road or your car being forced around the corner by a piece of road. In some senses you have an option of continuing on in a straight line, and you don't take that option because the road "forces" you to bend. Once one sees this as an issue, one sees it all the time: this may not take away Latour's main point, but it does make it less dramatic and makes you start to think more about the sort of situations that make it relevant, about what counts as having morality built into an artifact and about how you know when morality is built in. Because morality itself, for Latour, has to do some signification work, and that raises other issues. First of all the signification is subject to what we call "interpretive flexibility," which is part of the hermeneutic

---

is considered in terms of the ways in which it is present as a working material object—a quick and easy food-heater-upper—yet another role in its use context becomes visible: it appears to be able to change human eating habits. The microwave facilitates a particular kind of meal, the frozen, ready-made kind that can be 'prepared' in a short period of time and for a single person. It promotes such meals amongst its users, thus fostering a change in eating habits in which fewer are taken in company and more are eaten solo.

106. Cockburn and Ormrod 1993.
107. "We have been able to delegate to nonhumans not only force as we have known it for centuries but also values, duties, and ethics. It is because of this morality that we, humans, behave so ethically, no matter how weak and wicked we feel we are. The sum of morality does not only remain stable but increases enormously with the population of nonhumans. ... The impression given to those who are obsessed by human behavior that there is a missing mass of morality is due to the fact that they do not follow this path that leads from text to things and from things to texts" (Latour 1992, 232–233).

process: it depends on the discursive context. So how do you link that to materiality? When you pin down a meaning, how do you link it to that materiality? Because the materiality does not speak for itself.

So in that article I'm taking an example of a sign that has replaced a piece of materiality. Latour looks at a piece of materiality as a speed bump that has, in a way, replaced a sign, the sign being "slow down," the materiality is the speed bump. I look at the reverse. Let's take an example of where there used to be a dog fence, keeping a dog there, and now there's a sign saying "invisible fence."[108] How do we know what the link is between these two? So I quickly opened it up, problematizing it. What I am doing is just trying to ask questions about Latour's approach. Of course, I'm not saying that SCOT answers all questions, but SCOT, on some of the issues of the material infrastructure, where there are these political contestations over material infrastructure, is a very powerful approach for getting at the politics of technology.

ST   If I understand the examples you made, your objections are both methodological and epistemological: on one hand, delegations are too many to be described and we need a heuristic guide to select the ones that are relevant for analysis. On the other hand, what has been delegated, and to what such a thing has been delegated, is ascertained only through a hermeneutic process, and consequently prone to an interpretative flexibility that is removed from the analysis. If I should try to apply this last point to the speed bump example and maybe push it a bit further, I would say that it is true as a matter of fact that, when I see a traffic bump, I slow down with my car, but it is also true that it never happened to me to break an axle shaft on a speed bump. My knowledge of what will probably happen to my axle is playing a role: it makes me slow down. It is not so different from the example you made about the moon made of cheese, is it? The delegated morality and, thus, the "agency" of the nonhuman also depends on what I know or believe about my car and about the speed bump. So, unless I don't simply stumble on it—to quote Collins—we have once again a hermeneutic process that suggests us to move our analytical focus to the discursive context and to the actual actors involved.

108. An "invisible fence" is a technological device used, as traditional fences, to spatially restrain the mobility of dogs: a sound alarm signals when the dog is getting near the "fence," and it is followed by an increasing electric shock discouraging the animal from proceeding further.

TP Yes, exactly. There is a hermeneutic process. You cannot understand the speed bump out of its context. In that context there are laws: there are legal representations and there are social representations. There is this whole process of representation going on. You need that full context to understand how speed bumps "act." You can't just detach and say, simply, that there is a form of material agency given by these speed bumps. For me it is a much more complex issue. As you said, you have people recognizing that there are speed bumps, and the possible economic damage they could do to their cars. So there is also economy in there. It is just a trivial issue: if you were in a world where car owners had a huge amount of money, that speed bump would be much less of a constraint for you. You might even find it fun to smash your car every time: "Great, I'm going to get another car!" So there is an issue of political economy that is built into this. And there is an issue of representations and an issue of law. All these things tend to get effaced at looking at the material agency as the dominant thing.

ST Yet, several approaches in different disciplines are emphasizing how our relations with materiality may elude the representational level and the attribution of symbolic meanings. In particular, I am thinking about non-representational theories in human geography,[109] and their stress on affects, embodiment, and bodily habituation to the material environment. These approaches have now become very influential also in media studies, for example to address the role of media engagement in processes of place-making.[110] How does SCOT take on this challenge?

TP I am not familiar with nonrepresentational theories in human geography, but a notion of embodiment and of embodied knowledge is certainly relevant. As you know, SCOT is a part of a wider approach that had the concept of tacit knowledge from its very beginning. By definition tacit knowledge is knowledge that cannot be articulated: sometimes that knowledge would be the type of knowledge of positioning your body to learn to play a musical instrument, where your hands are doing something and you cannot articulate what it is. I don't see it as contradicting the attention for the representational level.

A part of knowledge can be articulated, and we use the representational idiom, saying what is the meaning of those things. Collins and I did a

---

109. See Thrift 2007, Anderson and Harrison 2010.
110. See Moores 2012, Tosoni 2015.

study with a vet student of mine, Larry Carbone, of veterinary surgery.[111] We discovered that the skills of the vet surgeons could be partly articulated via something we called the "hardness" of a skill. Neophyte vets have to learn things such as how long to search for an organ. In our case the veterinary surgeon—who was actually Larry Carbone—was desexing ferrets and he had to learn how long he should leave the animal open on the operating table before he concluded that he had searched long enough for a uterus to remove. In this case the ferrets were being spayed before being turned over as pets and there was uncertainty as to whether they had been spayed already. This criteria of the "hardness" of a skill links to all sorts of practical activities such as repair. How long will it take to repair a fridge? Skilled repair people can always give an estimate in advance. How long should you spend searching for software errors and so on? So the sort of bounds around skilled practices can be articulated to an extent and passed on and learnt as part of acquiring the skill.

A part of skilled practice cannot be articulated and we have a language for dealing with that part. We talk about it and there we use the language of "tacit knowledge." It is something that is learnt and communities have practiced. We cannot say much about it, but we can articulate what are the situations it is learnt in, or how it is passed on. You can talk about the social properties of that form of knowledge. I don't think it takes you out of the representational idiom. I mean: tacit knowledge is a special case. It is a strange sort of antiknowledge, since by definition it seems to be passed on without being articulated, but you can still describe its properties. Collins's book on tacit knowledge distinguishes between three different sorts:[112]

111. Pinch, Collins, and Carbone, 1996.
112. Collins 2011, 85–86:

Strong tacit knowledge, otherwise known as collective tacit knowledge (CTK) … is a kind of knowledge that we do not know how to make explicit. … Strong tacit knowledge … is the domain of knowledge that is located in society—it has to do with the way society is constituted. … Medium tacit knowledge, or somatic tacit knowledge (STK) … has to do with properties of individuals' bodies and brains as physical things. Weak, or relational, tacit knowledge (RTK) … is knowledge that could be made explicit … but is not made explicit for reasons that touch on no deep principles that have to do with either the nature and location of knowledge or the way humans are made. Collective tacit knowledge turns on the nature of the social, somatic tacit knowledge turns on the nature of the body, but relational tacit knowledge is just a matter of how particular people relate to each other—either because of their individual propensities or those they acquire from the local social groups to which they belong. Relational tacit knowledge turns on the way societies are organized.

social tacit knowledge, embodied knowledge that he calls "somatic," and what he calls "relational knowledge," which is knowledge that is tacit but only in certain circumstances. In principle it could be articulated but you choose not to, because it is more convenient not to. It is in relation to the situation, that's why he calls it relational. So I think that within the social constructivist approach there is a sophisticated language to describe the aspects you were mentioning.

ST   Going back to the other objection you raised, the methodological one, I have to say that I have experienced myself the problem of needing a supplementary heuristic guide when dealing with nonhumans through the lenses of actor-network theory. My attempt was to derive from the ANT's repertoire some sensitizing concepts that could help me to deal with the relationships between heterogeneous elements. This was for a project I was doing with a colleague and a good friend of mine on how media contributes to shape urban space.[113] We needed something to investigate the relationship between media practices, representations, and materiality, and I have to say that ANT offers very powerful tools for this: of course, "delegation" is one of them, but so, for example, are "translation" or "chains of translations,"[114] the ones we used. The problem is that, at least for urban space, once you are "sensitized" enough, it becomes impossible to "follow the network" of actants because the complexity of the relationships you start to see appears to be overwhelming. You really need a rule of selection among all these intertwined chains of translations, something to simplify this rhizome enough to describe it. And I think that SCOT, with its focus on relevant social actors, could play that role well. I hope this doesn't sound too outrageous to you!

Anyhow, I have to say that I have also some political concerns for this idea of morality delegated to nonhumans, or mediated by them.[115] I admit

---

113. Tosoni and Tarantino 2013.

114. "In its linguistic and material connotations, the concept refers to all the displacements through other actors whose mediation is indispensable for an action to occur. ... In place of a rigid opposition between context and content, chains of translation refer to the work through which actors modify, displace and translate their various and contradictory interests" (Latour 1999, 311). "[Translation] means displacement, drift, invention, mediation, the creation of a link that did not exist before and that to some degrees modifies the original two" (ibid., 179).

115. For a similar approach, see Verbeek 2006, 2011.

that reading Latour for me is always fascinating but is also always puzzling, since he continuously jumps in and out of a metaphorical rhetoric register that makes it difficult to understand when and to what extent he must be taken seriously. Latour himself honestly admitted at a later point that he "exaggerated" his positions on the issue of morality.[116] Edwin Sayes[117] has recently underlined this admission in a paper he published in *Social Studies of Science*. Furthermore, the same example of the speed bumps you quoted is rewritten in a book that came out in 1999, *Pandora's Hope*, and there Latour is definitely more cautious on this issue of morality. The fact is, anyhow, that when Latour talks about "morality" delegated to nonhumans or, if you want, about looking for the missing masses of our society in non-humans, in my opinion he is really risking embedding serious ideological implications into his approach: the idea that moral problems, or problems related to the social contract, can be solved through technological means. I really do not want to go into the problem of giving a rigorous definition of what morality is, but for sure it is not a definite set of constrains to action, prescriptions, or specific courses of actions. Morality or, say, social responsibility—I feel a bit uncomfortable when handling a concept like morality—are somehow at a metalevel in respect to specific prescriptions or courses of actions.

Maybe I can make my point with the speed bump example: you really would like people to be so responsible as to drive slower when they're passing in front of your house, and if they don't, you can force them to do so by embedding this prescription in a piece of technology. But nothing of morality, or of responsibility, or of a social pact between you and those people has been delegated to the nonhuman in this process. I mean, there's reason, and the reason why I slow down when I'm passing in front of your house.

TP What reason is that?

116. "We cannot infer from the current usage of terms of duty and authorization that technical objects have an obvious moral dignity in themselves. Therefore, that has not exactly been my intention. It is mainly the contempt that sociologists have for matter and for technological innovation which has led me previously to exaggerate somewhat in speaking about the 'tragic dilemmas of a safety belt'" (Latour and Venn 2002, 254).
117. Sayes 2014.

ST Well, I can slow down because I'm forced to slow down by an artifact, and I can slow down for a sense of social responsibility toward you. Social responsibility or morality is not a set of courses of actions whose execution can be granted by an artifact. There cannot be symmetry here because we are talking about things that are somehow at different levels: social responsibility is more a stance that can generate different courses of actions than a defined set of translatable courses of action. It is a principle to generate them: it is more similar to a *habitus*, in Bourdieu's sense. It is a stance that produces different kinds of actions.

TP Ah, I see. Yes.

ST The fact is that if a social contract between you and me exists, I will slow down when I am passing in front of your house, but I will also avoid honking if you are sleeping, or throwing the butt of my cigarette on your yard. If it doesn't exist, what are you going to do? Are you going to deploy another two pieces of technology?

TP Yeah, I see.

ST This seems to me a very persistent techno-utopia: the idea that you can solve social problems by addressing them as engineering problems, symmetrically. But this is not going to work because you cannot translate a generative stance into technological constraints. You can try to translate only some of the courses of action that derive from it. Lyon made this point in his works on control and surveillance technologies,[118] for example, but he is really just one of the many. I don't think we are so far from the problem of the "testing society" that we were discussing. In some respects, it's still the problem of the detachment of functions and meanings, or functions and values, and you can go back to Marcuse if you want. My objection is political, but in Latour's approach this detachment risks becoming embedded in methodology. It's implied in his provocative symmetric vocabulary. I understand Latour's Machiavellian stance, but this makes it more difficult to elaborate on a critical discourse with these conceptual tools, notwithstanding the intentions. Probably it is not coincidental that Latour had to correct himself exactly on this point. Or that, even in the amended version of his example on street bumps, he still claims that to an external observer it doesn't matter how the desired

118. Lyon 2002, 2003.

behavior—the car slowing down—has been obtained.[119] Ironically Latour also has a chancellor impersonating the external observer! Well, I guess it depends on the observer if it matters or not, if she is not an abstract observer. It matters to me for example; politically, I think it should. I hope I have made my opinion clear enough.

TP Yes. It seems like you had actually two criticisms, which I think I agree, on both of them. One was the social contract as a *habitus*, which is a lot more complex than a simple rule embedded in a piece of technology because all the commitment embedded in a technology is one simple rule: "Slow down at a certain speed." And in fact it's a nuance, it's more like a set of expectations that don't take necessarily deterministic form in normal human interaction. So that's one criticism, I think: that something complex—however you want to call it, form of life, *habitus*—has been reduced only to one rule. Which is, for a sociologist, a terrible thing to have in the world. I agree with that, that's a really good point. I actually haven't thought about that, but that's a good criticism. The second one is about the business of function and meaning that in this approach are detached. I think deliberately so: it is a leverage to say something. But, then in certain situations you may want to criticize technology in terms of its meaning, put the function and meaning back together, and it's harder to do. It may not be impossible, but I agree, it's harder to do with this sort of actor-network approach. I think I agree with that as well. The social construction of technology approach definitely gives you more political leverage when dealing with issues of function and meaning.

ST The differences between your take on materiality and the one advocated by actor-network theory are further discussed in another paper we have already mentioned: the one you co-authored with Asaf Darr[120] on the social organization of obligation in selling. There you advance a dramaturgical and material approach to selling, and you compare your own concept of "material script" to Akrich's and Latour's notion of "script."

119. "The driver modifies his behavior through the mediation of the speed bump: he falls back from morality to force. But from an observer's point of view it does not matter through which channel a given behavior is attained. From her window the chancellor sees that cars are slowing down, respecting her injunction, and for her that is enough" (Latour 1999, 186).
120. Darr and Pinch 2013.

TP   Right, in that paper I have addressed the exact difference between what we call "material scripts" and what Madeleine Akrich[121] calls a "script."[122] The key difference is this: obviously materiality plays a role, but for our definition of "material script"[123] the context of accomplishing a task is the most important thing in how this materiality works.

The example I can give you is a shopping bag. A plastic shopping bag has many material scripts or affordances—if you want to use this term—built into it. It can be used for carrying stuff, but it can also be used to be put over your head in a Halloween mask. If it is a black bag you can cut it up and just use the plastic material to put over your window, if you want to

---

121. Akrich 1992b.

122. "When technologists define the characteristics of their objects, they necessarily make hypotheses about the entities that make up the world into which the object is to be inserted. Designers thus define actors with specific tastes, competences, motives, aspirations, political prejudices, and the rest, and they assume that morality, technology, science, and economy will evolve in particular ways. A large part of the work of innovation is that of 'inscribing' this vision of (or prediction about) the world in the technical content of the new object. I will call the end product of this work a 'script' or a 'scenario'" (Akrich 1992b, 207–208). See also Akrich and Latour 1992, 259–260:

Script, description, inscription, or transcription: The aim of the academic written analysis of a setting is to put on paper the text of what the various actors in the setting are doing to one another; the de-scription, usually by the analyst, is the opposite movement of the in-scription by the engineer, inventor, manufacturer, or designer (or scribe, or scripter to use Barthes's neologism); for instance, the heavy keys of hotels are de-scribed by the following text DO NOT FORGET TO BRING THE KEYS BACK TO THE FRONT DESK, the in-scription being: TRANSLATE the message above by HEAVY WEIGHTS ATTACHED TO KEYS TO FORCE CLIENTS TO BE REMINDED TO BRING BACK THE KEYS TO THE FRONT DESK. ... No description of a setting is possible or even thinkable without the mediation of a trial; without a trial and a crisis we cannot even decide if there is a setting or not and still less how many parts it contains.

123. Darr and Pinch 2013, 1612:

A "material script" can be defined as a recurrent pattern of interaction involving material objects in a specific setting. It is important to note that this definition of a material script differs from the use of "script" for dealing with material objects found in actor network theory by Akrich (1992) and Latour (1992). In actor network theory scripts are more restrictive, they are semiotic embeddings (reflecting the imagination of the designer) within a material artefact—typically a piece of technology—of a program of action for a user to undertake in interacting with the artefact. ... Material scripts, although they may be embedded in material artefacts, depend for their interpretation upon the specific ... context—what we might call the setting—in which the artefact is used. Material scripts require that humans are able to recognize the appropriate courses of action and the obligations entailed. In short, for us a material script means an appeal to a commonly shared understanding of the use of the material object in a particular setting.

blacken it out. It can be used in many, many different ways. What we say is that a material script is only relevant for the particular social context of accomplishing a task. The example we give is from pitching. The pitchers have a routine called "bag nailing," where at a certain point in a sale they hand shopping bags to the customers.[124] They hand them out. One, two, three, maybe ten, twenty bags. This is before the people know what the final price of the goods, say a set of perfumes, are. They take the bags and the bags somehow obligate them to purchase the goods. When the final price is announced, the good would be placed in the bag. People always take the goods once they take the bag. There is a kind of material script that is built in the scene, the bag obligating you to purchase the goods, even though you have not yet technically purchased the goods.

It happens something similar with the materiality of the supermarket shopping cart: it puts you under a weak obligation to purchase the goods you put in the cart even if you haven't legally purchased them.[125] Materiality plays a role there: this is why this thing is designed to be so deep, and it is so hard to take the stuff out. It is something that you know you are not supposed to do: you sometimes see someone slyly leave an article on the shelf hidden from the view of the cashier. It would be very strange if you arrived with your shopping cart at the cash point and took every item out so as not to pay. Basically you are placed under this obligation by the materiality of the shopping cart and by this material arrangement.

With the pitcher example the interesting thing is that the goods are not even in the bag, you are just holding the bag. Even just holding that bag in that setting creates obligation. So the difference between Akrich's definition

124. "By giving out carrier bags to people [who have shown an interest in the goods], the display of interest (hand-raising) is transformed into a more tangible and thus more ineluctable obligation to buy. The acceptance of a bag ...—that is, something that ordinarily would be used to carry away the goods—implies consent to the imminent transfer of ownership of those goods" (Clark and Pinch 1995, 100).
125. "The appropriate material script of a well-laden supermarket trolley as it is wheeled towards the checkout in the supermarket is as a container for conveying goods shortly to be purchased, and the accompanying moral script is that of a good consumer who ... is under a weak obligation to purchase those goods in the trolley. But a supermarket trolley full of objects being wheeled in other contexts has a very different material and moral script. For instance, when being wheeled by a homeless person on the streets it does not mean the objects in it are to be purchased and signifies a very different moral script" (Darr and Pinch 2013, 1612–1613).

of script and ours is that for her the script seems somehow put in by the designer and it is fixed. Actually she is more subtle than that, because she has the notion of reinscription.[126] But she is often read, as Latour is often read, in a deterministic way, like it is just the technology and the materiality that are doing something. We wanted to stress instead that it is not just that: it is in a particular context of social action that this thing has been accomplished. Otherwise you would have a rather deterministic reading of scripts. This is the essence of the difference between the way we are using "material script" and Akrich is using the term.

ST On this same topic, scripts, you have coauthored with Paula Jarzabkowski[127] another paper where you propose what you call an "accomplishing approach" to sociomateriality. How does this approach relate to the notion of material scripts?

TP That later paper further clarifies some of the differences between the Latour-Akrich notion of scripts and the notion of material scripts outlined in Darr and Pinch. It also offers a broader agenda for doing sociomaterial research in social sciences and in management studies, since the paper came out on a management studies journal. Here the critique of Latour and Akrich is that they tend to deal with sequences of action which are pre-scripted and they do not pay enough attention to the need to look at how materiality figures in the ongoing accomplishment of interaction, including the wider setting such as norms, legal constraints, and other aspects of socioeconomic life.

As I have already explained, the particular economic obligation contained in the shopping bag only becomes apparent in the accomplishment of the pitching interaction. We could also call this the performativity of material scripts. I sometimes call this whole way of looking at things "material performativity." Material scripts have to be performed or accomplished in the unfolding of ongoing interaction. Essentially, in the new paper we deal with an example of everyday interaction with material objects where

---

126. "Re-inscription: The same thing as inscription but seen as a movement, as a feedback mechanism; it is the redistribution of all the other variables in order for a setting to cope with contradictory demands of many antiprograms ...; the choices made for the re-inscription defines [sic] the drama, the suspense, the emplotment of a setting" (Akrich and Latour 1992, 262).

127. Jarzabkowski and Pinch 2013.

there has been a breach or disruption in the normal scripted interaction between a person and a material object.

The example discussed is that of a "limp clip card"—the clip card is a way of prepaying for multiple rides on trams and trains in some European countries such as the Netherlands and Denmark. Usually you insert your prepaid card in a machine before the ride and it will "clip" off the ride you have prepaid. But often these clips or strip cards get damaged in wallets and become limp, and the machine will not be able to read them or clip them, meaning you cannot use them to prepay for your journey. You in effect "lose" the ticket for that ride. This happened to me once in Copenhagen on the way to the airport with my last prepaid ride. It is very annoying! So I tell the story of how I was able to "repair" the clip card. A woman who I had helped into the elevator with her pram showed me how to spit on the damaged card and restore it to working! This is a case of accomplishing social interaction with materiality in a specific context and I discuss the full richness of how this repair came about—including the many materialities involved: the card, the card reading machine, the spit, the pram, the eleva- tor, the train—and the gendered relationship with the woman.

The point of the paper is to show that an accomplishing—or performa- tivity—approach takes us beyond the routine behaviors implicit in material scripts and their assumed sequences of action, which may make us blind to the possible improvisations until a breach, absence, or disruption occurs that occasions the repairs that make habitual action visible. For example, it is hard to see this culturally situated activity in the clip card until we see how its assumed properties and scripts are breached and repaired. Until then, it is just a ticket system, which is too mundane for us to notice how it is entangled with more fundamentally situated, cultural, political, and socioeconomic factors, such as our assumptions about the efficiency and economic exchange involved in travel.

To extrapolate further, we are often blind to our contextually bound assumptions about materials. If we do not see the multiple possibilities that go beyond our usual conceptions of, say, a bag, then what other criti- cal possibilities do we also miss in, for example, the gendered, racial or socioeconomic nature of materials and scripts that carry assumptions of inequality and division? The paper argues that if we are to make sociomate- riality an important and central agenda in social science research, we must enable materials to take center stage in situated activity and explore the

assumptions and social obligations that surround them, keep them working, and, if need be, enable change. So this is relevant also under a political
point of view.

3.6 Back to the Golem: SCOT and Politics

ST Talking about politics, a criticism that has been raised several times
against SCOT is that it tends to be politically neutral, if not blind. I could
quote, for example, Andrew Feenberg[128] for the case study on the explosion of the *Challenger* included in *Golem at Large*, or Stuart Russel[129] on the
early formulation of SCOT, or again Langdon Winner,[130] who at a later
point violently attacked the whole constructivist field in the sociology of
technology.

128. Feenberg 2006, 723:

It is important that the principle of symmetry not be invoked by researchers in cases where real
world asymmetries significantly bias outcomes. I would argue that just such violations are all too
common where technology is concerned and, worse yet, often several communities with different standards of judgment confront each other and vie politically for control. I do not see how
it is possible to sort out such technological controversies with a principle of symmetry originally
designed to understand much simpler scientific disagreements where adversaries are more nearly
equal in power and standards shared by all. Constructivists originally introduced the principle to
demystify Whig history of science and the technocratic pretensions it supports, but today cyni-
cal appeals to symmetry now excuse inaction on global warming and other controversial issues.

The authors replied in Collins and Pinch 2007.

129. Russell 1986, 332–333:

It is easy to slip from relativism as a method into relativism as a position on conflicting view-
points—what we might distinguish as "methodological" and "substantive" relativism. It is not
clear what Pinch and Bijker intend "technological knowledge" or "technical meaning" to cover.
But whatever definitions they may wish to take, arguments about technical performance blend
into arguments about objectives, economics, effects, risks, benefits and other dimensions of their
desirability. If we accept that arguments over technological options are socially constructed, then
it follows that a relativist approach with respect to them leads us into relativism with respect to
social interests—in other words, political neutrality.

The authors replied in Pinch and Bijker 1986.

130. "Although the social constructivists have opened the black box and shown a
colorful array of social actors, processes, and images therein, the box they reveal
is still a remarkably hollow one. ... They offer no judgment on what it all means,
other than to notice that some technological projects succeed and others fail, that
new forms of power arise and other forms decline. ... As compared to any of the
major philosophical discussions of technology, there is something very important
missing here, namely, a general position on the social and technological patterns

TP Yeah. But I think the claim that there's no politics in SCOT is silly. SCOT gives you a technique with which to examine technology, unpack it, and criticize it.

ST It seems to me that these criticisms derive from an idea of the political role of the sociologist that is very different from yours. SCOT has in fact a very clear and specific take on politics: I would say that it doesn't aim to replace a discourse over technology with another. On the contrary, it aims to open up a discursive field and to enable social actors to produce their own discourses. To tell people "how to think," and not what to think, about technology, to quote the subtitle of *Dr. Golem.*

TP Yeah, absolutely. That's seen as a weakness by people who want to condemn, like Langdon Winner, some sort of technology because somehow they have already decided in advance that certain technology is repressive. SCOT doesn't make that move immediately. I think it could do it in the long term, enter into that sort of take on technology. Nuclear power, for example: SCOT does not say, "Nuclear power is bad." It says, "Let's examine the particular artifact of nuclear power in a social context, and then we'll find out what it does, what meaning it has, possibly what good it does, what harm it does, for which social groups." But it's not a predetermined issue. This is opening the black box, but for some people like Langdon Winner, that's bad because they want to make a statement in the end where you condemn a technology. For them that is not politically strident enough. It's too nuanced.

ST I guess that the problem is once again symmetry.

TP Yes. I think that the criticisms we have received from Andrew Feenberg are a very good illustration of this whole point, of what counts as having a politics. Andrew Feenberg is a very well-known philosopher of technology. He was a student of Herbert Marcuse and is a very good guy—a friend of mine. He has a very sophisticated take on technology. I don't want to imply that somehow I don't find his works sophisticated. But we had this debate with him about the analysis of the space shuttle *Challenger* accident. He uses us as an example of why a symmetrical

under study" (Winner 1993, 374–375). For an answer by Pinch, see Pinch 1996. For a direct debate on the topic between Pinch and Winner, see the footage of the STS 20+20 Conference at Harvard University, Apr. 7–9, 2011, http://www.youtube.com/watch?v=D9o2B47CArw (last accessed Apr. 8, 2016).

approach to that accident is wrong because it doesn't give you enough political grounds to say more. He wants to make strong statements about how NASA was kind of irresponsible. I think that is what he would have liked to say. He liked to do more pointing blame, and symmetry does not allow you to point blame in that easy way. In response, we pointed out that there's a politics attached. Even adopting the symmetrical approach to analyze a major technological accident like that, you can say things about the politics of technology.

Here's one of the things we actually say in *The Golem at Large* about the accident: we actually criticized NASA for presenting the space shuttle as this kind of technology that's like an airplane, and that has led them to the mistake of flying civilians. This is why we had a school teacher[131] tragically killed in the *Challenger* accident. Amazingly, Diane Vaughan pointed out to me that NASA was about to reinstigate the Teachers in Space Program again, with the *Columbia*, just before the Columbia accident happened.[132] Diane Vaughan was on the *Columbia* accident enquiry, and she wrote a great book,[133] which we used in *The Golem at Large* on the *Challenger*'s launch decision. This thing is amazing: why are they flying civilians in a dangerous, uncertain technology? Because somehow they have fooled themselves: "It's like a 747, it's a pretty reliable technology." So we say that a politics of technology here could involve just presenting something as it really is: it is an uncertain technology, with lots of uncertainties, and we shouldn't hide it up. That's a politics of technology that you can draw from our article. It says something that can change the world, the sort of people who can fly in this: it is also a politics. And so there's a politics even in the symmetrical analysis of accidents. That is what that debate was about. I just liked that debate because it brought up again the issue of symmetry as applied to technology.

ST Just to recall the main points of the debate, you disagreed about who was to blame for the explosion of the shuttle. Feenberg was among the many who thought that NASA managers had to be held responsible for the

131. The schoolteacher Sharon Christa Corrigan McAuliffe (1948–1986) was one of the seven crew members who died in the *Challenger* disaster; she would have been the first teacher in space.
132. The *Columbia* disintegrated while it was reentering the atmosphere on Feb. 1, 2003.
133. Vaughan 1996.

incident because there had been several alarms about O-rings' resilience at low temperatures. There had also been this dramatic teleconference in the evening before the launch where the engineers of Thiokol, the company that produced the solid rocket boosters, advocated a postponement of the launch. NASA managers decided not to delay the launch of the shuttle any-more. In your paper you claimed that NASA managers had reasons not to postpone the launch.

TP Yeah, NASA had good technical reasons. Within its own framework of what sort of judgments had to be made about what data are reliable data, it had good reasons to launch the space shuttle that night. It was in hindsight the wrong decision, but it was made for good reasons in the institution. Maybe it's the nature of that technology that people act with the best of reasons and accidents still happen. It is impossible to prevent accidents. It's an amazingly complicated machine and it's rather unreliable.

ST One of your key argumentations was that you cannot operate an unre-liable technology like the shuttle without constantly having reasons for concern about the malfunctioning of one or another piece of this very com-plex technological system. It is up to the community of experts to decide which of these reasons of concern can be ignored and which of them can-not. If they wouldn't decide that some of them can be ignored, they would never operate the technology. The principle of symmetry in the case study of the shuttle led you to methodologically ignore that the decision taken that day by NASA was wrong and to analyze the reasons that drove NASA to make that decision in its own context: when the decision was made, and not in hindsight. In disagreement with the general criticisms raised against NASA managers, you showed that there were good reasons. Public opinion, but also many philosophers and sociologists of technology, were blaming the wrong people for the wrong reasons. The blame, for you, belonged to whoever convinced the public, and maybe themselves, that that technol-ogy was safe.

TP Exactly.

ST Since we mentioned Lyon at an earlier point in the interview, I'd say that the case of the shuttle reminds me of the criticisms of the US government about the intelligence warnings that were supposedly ignored before 9/11: a kind of keystone in all the conspiracy theories on the issue. People seem to switch from a complete trust in the efficiency of the sociotechnical system, responsible for intelligence and security, to conspiracy theories. This happens because a particular technology, in this case the intelligence and security sociotechnical system, has been sold to them as efficient and secure—and as a way to cope with sociopolitical problems, the ones from which dramatic security issues derive, through technological systems.

TP Exactly. These are all examples of flip-flop thinking. We have already mentioned this earlier in the interview: you have switches that are called flip-flops, and they switch from one state to another in integrated circuits. One version of science and technology is all about certainty, and then, when people get somehow disillusioned, they flip to the other view, where: "Scientists are not genuine," "They are not genuinely trying to do their best at their jobs," "They put lives at risk for their own selfish interests." Somehow they would start to evoke conspiracy theories, like with the government case. They start to believe in some incredible conspiracy, that 9/11 was actually a conspiracy by the government: it is the next twist of this sort of argument. But if you had a view of how these complicated systems are—social systems, which are also, as you nicely pointed out, technological in character, like information gathering systems—that would not happen. If you don't invest overweening trust in them, people won't be surprised when occasionally things go wrong or they're messy. They wouldn't flip, when disappointed, to the opposite viewpoint. Sure, you can still criticize managers or engineers for their decisions and do things about them: NASA has taken lessons from that accident. But somehow to reject the whole system is kind of the wrong approach.

ST Of course, building this kind of faith in unreliable sociotechnological systems is also a political strategy for their deployment. And not just for that: it's a political resource that can be spent in many ways. It seems to me that a deconstructive take on this kind of faith, when it is needed, is a precious political contribution we can expect from SCOT and from its principle of symmetry.

TP It's what we hope for in the *Golem* books: helping citizens make decisions in a technological society in a more sophisticated way.

ST As we said, SCOT and the *Golem* books aim to tell people how to think, not what to think, about technology, science, and medicine. Under this point of view, the most political chapter of the whole *Golem* trilogy, the one that makes your take on politics more clear is, in my opinion, the one about vaccination in *Dr. Golem*.[134] In that chapter you have two distinct voices: they are somehow "real" voices, since there's you and your wife Christine in the role of the interviewee, arguing with Collins in the role of an aggressive interviewer. Nonetheless, the final result is not that far from those reflexive rhetorical devices deploying two or more voices throughout the text. The paper uses this strategy to discuss different points of view about vaccination policies inside the discursive field that your approach has opened. We have the Pinches in the role of the informed parents defending their decision to refuse to administer to their first daughter the standard cocktail of vaccines used in the United States at that time. They wanted to substitute the whooping cough vaccine with the inactive version, known as the DAPT, commonly used in Europe. Collins, on the other hand, tries to demonstrate that the Pinches' concerns about possible side effects are not solidly grounded. He points out their insignificant statistical incidence and the uncertain information about them. This finally drives him to accuse the Pinches of giving priority to their own family interest over social interest.

TP Yeah, it enables the politics between the two of us, in this case, to come forward and to be articulated. But we only found that by accident, we were in a big fight. Collins wanted to impose a particular "technocratic" version of the sociology of medicine on me! I was kind of refusing it. In fact, I've refused what he wrote there even more in another, unpublished paper I wrote for a talk. But I think I'll bear in mind what you said, that is the most political chapter in *Dr. Golem*. I think that's a good insight. Actually I take another turn on the political issue in this other paper by ending up slightly more hostile to Collins's view than I was in that book, for these reasons.

ST Actually, it's the whole book, *Dr. Golem*, that seems to me the work of yours where your take on politics is more clear and better articulated in all its passages and implications. I think we have at least four main points there. First of all, we have this initial key move that consists of opening up a discursive field: a move that on specific occasions requires the deconstruction of the technocratic claims that keep the field closed. In the book, you

134. Collins and Pinch 2005.

do this through several means, but a key role is played by your discussion of the placebo effect[135] in the very first chapter.

TP "The hole in the heart of medicine," we called it.

ST After performing this breach, as a second point, you carefully avoid closing right away the discursive field you have just opened, as would happen through the imposition of a specific point of view on the issue you are debating. This is the thing that comes out very clearly in the "Vaccination and Parents' Rights" chapter, thanks to the presence of two conflicting voices. It doesn't matter, in my opinion, if that chapter took that form by accident: what counts is that you decided to keep it in that way, showing how, in the discursive field that has been opened, there's still enough space for two conflictive opinions, both of them well informed and competent, confronting each other. Once again, everything is more related to "how to think" than to "what to think."

TP Yes, it's true.

ST Then, there's a third relevant point: this effort of opening a discursive field must not be confused with antiscientific positions or with the populist idea that everybody, however their expertise or competence, should be entitled to equally participate in issues of public policies regarding scientific, technological, or medical issues. The typical example here are the misinformed "awareness" campaigns, so typical of the Internet, that from time to time surface also in traditional media and even in the agendas of the most populist political parties. Actually, you stress how things are more complicated; Collins, in his work, has devoted a lot of attention to this delicate issue,[136] trying to clarify what the role of experts should be in a democratic policing process involving technological issues. There's another chapter of the book that is very illuminating about this: I'm thinking about

135. "The placebo effect—the Latin is placere, to please—is the name given to the alleviation afforded by the administration of drugs and treatments that have no direct effect on physiology; fake drugs and treatments often cure just as well (or badly) as the real thing for reasons that are not much understood beyond the phrase 'mind-body interaction.' The placebo effect shows that, at best, medical science has only partial control over its subject matter. The placebo effect, then, can make people better, but, at the same time, it is a massive embarrassment to the science of medicine. And that illustrates the main theme in a nutshell!" (Collins and Pinch 2005, 3).
136. See Collins and Evans 2002, 2007; Collins, Weinel, and Evans 2010; Collins and Weinel 2011.

the chapter readapted from the previous *Golem*, "The AIDS Activists," the one based on a book by Steven Epstein.[137] What strikes me most in this chapter is not only the fact that these activists were able to modify the way clinically controlled trials were conducted. It is also the fact that the scientific expertise they gained in the process drove them to acquire an understanding of the reasons and the needs of their counterparts, the doctors, and to modify their initial goals: sometimes, up to the point of receiving accusations of betrayal by their own community, like they had gone native in the process.

TP   Yeah, in "The AIDS Activists," they become like "positivists" at some point. Because they are taking on the medical scientific viewpoint. Yeah, that's a really great example.

ST   And finally, there is a last point: and in my opinion, this is the one with the most radical implications from a political point of view. I don't think you have ever done a similarly radical claim in any of your other works. The chapter I am thinking about is the one about the "bogus doctors"; ordinary (but not unskilled) people that pass for doctors. That chapter that comes immediately after the one about "the hole in the heart of medicine." The fact is that, in the chapter about AIDS activists, you and Collins clarified how medical expertise keeps playing a key role in the discursive field that has been opened once the principle of authority connected to an infallible "scientific medical method" has been deconstructed. In the chapter about bogus doctors, on the other hand, you show how the medical institution cannot be considered the only guarantor of "expertise" since it is not *always* able to effectively distinguish experts from nonexperts within itself. You even remark that when bogus doctors get unmasked, it's almost always because they bring attention to themselves for issues not directly related to their medical practices.[138] I do not want to push this argument too far: certainly not up to delegitimizing the medical institution. But, for sure, what I get is that the medical institution is not infallible in its task of ratifying expertise. It may have problems with that. Since expertise has to play a key role in public policy, this statement has very relevant political implications.

137. See Epstein 1996.
138. "Bogus doctors as are found out are nearly always uncovered not when they carry out some medical procedure incompetently, but when they fail in another aspect of their lives not directly to do with medicine" (Collins and Pinch 2005, 11).

TP Now, that's very interesting. You pointed out many things. First of all, I agree with the hole in the heart of medicine, and I think you're very astute about the politics we take on. But I think this extra claim that you make, and you've read into the bogus doctor chapter, as much as I'd like to endorse it, I'm not certain you'll find us saying that there.

ST You are right: I don't think you develop this point explicitly.

TP I don't think we actually say that this shows that medicine has a problem, or that this chapter is the most radical because it shows that the very mechanisms in medicine for stamping, legitimizing expertise, or deciding on expertise as well, somehow are failing. I don't think that we go that far. We probably take it as an empirical fact. Because in a way, it's probably true.

ST You don't make that point explicitly, but well: isn't it enough to present that empirical fact?

TP Well. Actually, you may have some reasons: I haven't really thought about this. It's probably true that institutions, all institutions, are like this. That they can't actually demarcate who's going to become the expert. I mean, every institution: rock music has this problem. Every scholarship, every institution must have this problem. There are probably lots of bogus experts. It's odd, you don't think of bogus rock musicians, but there probably are people who use that category and say, "These are not really musicians." So I think, I suppose, that every institution in theory has this problem. Still, I am not certain of the political relevance of this statement. In other words, it's an endemic problem of anything that tends to draw boundaries. Anything that draws boundaries has always had—seems to me—cases of things that fall out of boundaries and people who claim something falls outside the boundaries. There will always be this issues about who's got the expertise or not.

I think what's interesting in that chapter is that we showed the sort of expertise. Maybe it's what you are thinking, that the sort of expertise these bogus doctors have is actually one of the hardest ones to throw them out on because it's the sort of day-to-day routinization of medicine where they're so good. In other words, the stuff like the succor of medicine, the caring side of it: they can even be better at it than the people who have the legitimate expertise, and they can be very good on day-to-day practices. But I think we have a critique of them: there are moments in medicine where expertise really does matter. So you think about this: if you are having some health issues. You know, suppose you have got some rare condition, some rare cancer, and one form of treatment is radical chemotherapy that is probably going to kill you. And this diagnosis is a very rare case where, in fact, there's a chance of some other drug that may help you. You would really depend upon the expert, you know, to make that diagnosis, and you would care about it. It's in those sorts of rare cases that I think that the expertise matters.

Although you can read a political critique into it, okay? You can say it's more radical. But in some sense I think we're probably saying that, for the good of medicine as a whole, then you want be able to tell who the genuine experts are from the bogus experts. So it's politically conservative in that sense, I think; ultimately, I think that most people, and we would include here us as analysts, would want to say that ultimately it does matter and you want to be able to tell real medicine. So the message isn't that medicine should abandon policing its boundaries. It still should do that. So there's not an undermining of medicine, like an Ivan Illich argument, at all.

ST Yeah. I understand this, and also I totally agree with your point of view: it should go on guarding its boundaries and finding better and better ways to do that. You indicate, for example, how the recognition of the category of "paramedics" is a way in which the institution redefines its boundaries to better cope with this problem of "social" expertise.[139] So, it's not an undermining of medicine that I think I can read in your book, nor I am advocating it. What I want to say is this: pointing your finger at the "hole in the heart of medicine," you open a discursive field. In this discursive field, expertise plays a key role. But the institution can be unable to ascertain where the expertise is—or at least, a specific kind of expertise—on an a priori basis in the same way that it cannot easily tell bogus doctors from official doctors. This means that this expertise, that plays such a relevant role within the discursive field that has been opened, must be ascertained within the discursive field itself. I agree that within a specific discursive field, we can agree that an expertise in the day-to-day routinization of

139. "Nowadays the circle is being squared by the recognition of the medical efficacy of less qualified groups. Nurses are now allowed to be seen taking on increasing responsibility, while the new category of 'paramedic' recognizes how much can be done in the absence of a complete medical training" (Collins and Pinch 2005, 59).

medicine, or the succor of medicine as you call it, doesn't have to count that much. But the fact is that you cannot infallibly rely on an external institution closing again the discursive field by authority: by an external and official authority. And this, because—as you have shown—this authority may be not infallible in defending its boundaries, in deciding who has expertise and who does not.

In my opinion, the example of rock is a bit misleading, in this context, because you don't have an institution that has the duty and the power to legitimate, with a specific and well-defined process—and with a ritual— who is officially inside it and who is outside: who really has expertise. Especially nowadays, the music institution works in a way that is far more complicated: there are parts of this world, like big labels and institutional journalists, that work on a top-down basis and other parts that work on a bottom-up basis, from do-it-yourself productions to consumers' choices. And you also have a very wide, blurred category in between, from semi-professional bloggers to small labels. So you have many different processes of legitimization and not just an official one. There's no political stake in demonstrating that there's not an official and infallible authority to guarantee expertise in this field. Even if I have to say I've found your paper about reputation on the Internet[140] very intriguing since it contributes to demystify these rhetorics about "democratic bottom-up systems" that are so typical of the new medium.

TP   True. About the music example, you are right: there's not that kind of legitimizing authority. Or, anyhow, it's very weak.

ST   But for medicine, things are very different: you are deconstructing something that is far more relevant there. It's sure that when I go to a doctor, I expect him to be a doctor, and I want him to be a legitimate doctor. But this case of the bogus doctors makes the situation a bit more blurred and becomes relevant when a discursive field gets opened—for example, in a controversy. Politically, it's not enough to open the black box of medicine if you don't open the black box of the process of legitimization of expertise as well: otherwise, the discursive field that was kept closed by a blind faith in the scientific method in medicine will be kept equally closed by a blind faith in expertise institutionally legitimated. This is what you do with the bogus doctors chapter: that's why I hold it as the most radical. But this

---

140. David and Pinch 2006; see also Pinch and Kesler 2011.

doesn't imply any antiscientific or delegitimizing stance: on the contrary, in the case of the AIDS activist, after a harsh confrontation, what happens within the discursive field is a reciprocal acknowledgment of the specific kinds of expertise of doctors and of activists. That process changes both the social actors. I see the chapter on bogus doctors as the *pars destruens* of the take on politics of SCOT, or at least of what I think is its take on politics, and the chapter about AIDS activists as its *pars construens*.

TP Yeah, that's a good point. I think I agree with that. It opens up the discursive field. You can start to think about that. This may have other implications: it also could be relevant in a situation with scarce resources, where you would be really glad to have bogus doctors to actually be practicing if they wouldn't be allowed to make decisions on really important cases, like those involving rare conditions. That would be a radical position because medicine would never allow it. It's then a matter of resources and politics, what's available. But anyhow, yes, it opens up: I can agree with you. It is actually a radical critique, I haven't thought of it. I am glad you're writing this book instead of someone else because you're politically more perceptive about *Dr. Golem* than we are!

ST Well, I am just pointing out something that was already there, or at least that I think I can see in it. There's a last point you make in *Dr. Golem* that you have somehow already mentioned and that I'd like to discuss: you talk about a sort of misalignment between the personal level and the social level of medicine. Between, if I got it right, what Foucault would call "biopolitics": the good of society as a whole regarding health and the level of personal good.

TP Oh, yeah: that's a point that I think we've got very early on, and you can see it in the context of the clinical control trials of, say, AIDS drugs, where it's to the advantage of the individual in a clinical control trial, who has a deadly disease like AIDS before they'd got any drugs for AIDS, to mix up the placebo and the drug limb because the individual might benefit. This benefits the individual if the drug actually works. There's no guarantee, but it actually may benefit the individual, while the institution of science, medicine as science, is damaged because now you can't effectively say if that agent is successful or not, because it's messed up. The placebo and drug limbs of the trial are not clearly distinguished anymore. So there's an example where it's good for the individual but bad for the institution.

We have used those sorts of examples: vaccination, of course, with the free rider problem is a much better-known example of this. So we try and make a distinction between what can be good for an individual and what is good for the institution.

Alternative medicines are a very good example here; there is lots of weird and wonderful stuff in alternative medicine that has never been tested, never mind vouched for, by medical science as currently understood. We don't say an individual shouldn't do that.[141] If you know you are very, very ill, and you find that something is helping you: why not? But there would be a huge shift for the institutions of the science and medicine to invest in the idea that these alternative medicines are going to cure lots and lots of people.

ST Considering also that investing in a direction means not investing in another.

TP Exactly. So it just seems to us that there's kind of a distinction between the individual good and the social good. A lot of people haven't really thought about it too much, it seems. Some sociologists have, but there is a lot of confusion over these issues, so we try to clarify what that issue is about. We used the term "succor," an old-fashioned term, for this sort of good of medicine for individuals. You know: you want everyone to have as much succor as possible, but that is a different underside of science and medicine, the caring aspect of medicine, you as an individual being looked after. The other is the wider side of medicine. So you can frame it with Foucault, but I think that's a familiar distinction. I think there are three important points in *Dr. Golem*; first was succor, social good differentiated from individual good; second, the idea, which is an old idea, that medicine is an art of uncertainty. This is the message of all the other *Golems*, and of *Dr. Golem* as well: that medicine has moments of uncertainty. This is fairly obvious. That's why you have the phrase "the art of medicine": no one really believes that medicine is hard science. Most people realize there's

141. "Medicine as science and as a collective responsibility must not be driven by popular opinion, even though individuals may be right to seek out any unproven alternative for themselves when the ultimate questions are asked. To say 'leave it to the people' is to risk the abrogation of our long-term collective responsibility to scientific medicine, even if sick and dying individuals might still be wise to try the cure" (Collins and Pinch 2005, 110–111).

some uncertainty. So that's not such an important message for medicine because it's known already. The third point is about the politics of expertise, to which you've already alluded. Essentially the point is this: that a lot of people think of expertise only in terms of credentialed expertise. *Dr. Golem* points out that you can become expert in something without having the credentials. And then you have to get into what is expertise, the politics of expertise; that's what we've already talked about. So these are the three main issues.

4 Other Entanglements

4.1 Selling Revisited

Simone Tosoni   In which direction are you developing your work, at the moment?

Trevor Pinch   Well, that is not an easy question because as usual I am working on many different things at the same time. I would say that I have a renewed interest in selling, and that after my work on the Moog I am interested in looking deeper into issues concerning sound, so sound studies. In both cases, I am reemphasizing the theme of materiality, which is the link between my work on selling and my work on sound.

ST   I would start from your renewed interest in selling, before proceeding to sound studies.

TP   Okay. Actually I think there is a new, broader, interest in selling. Recently I have visited the Centre de Sociologie de l'Innovation in Paris and its new head, Alexandre Mallard, told me he has edited a book[1] on selling techniques, where he says he uses the work that Clark and I did in the 1980s. More or less in the same period, I have received a paper from this good sociologist of organizations in Stanford, Steve Barley, on car sales.[2] He compares car sales in the showroom to online selling and he uses the paper I published on Goffman and materiality.[3] It is so nice when a major sociologist sends you a paper on a topic you have worked on: so it wasn't crazy to work on selling! This guy suddenly says: "Yeah, why the hell is nobody studying selling?" We made all these arguments back in 1980s, looking at

1. Kessous and Mallard 2014.
2. See Barley 2015.
3. Pinch 2010b.

the sociology of work. The task of selling is so huge, if you just look at the number of people involved in the sales force in some form. Somehow everyone wants to talk about marketing and other exciting things such as branding, but routine selling is what most of these people are doing. They are not doing marketing or branding, or that exciting trendy stuff. They are doing routine selling! So this is work that we did many years ago that suddenly has surfaced up again. There is a bubble of interest in selling, which was so important also in my book on synthesizers. I suddenly now see much more interest in this strand of my work, and in particular in the materiality of selling.

As you know, I have been interested in studying selling ethnographically since the mid-1980s, when I did this work at York with Colin Clark, which led to *The Hard Sell* and to several articles. But I have to admit that at that time that book wasn't really taken up. Our approach was coming from conversation analysis and ethnomethodology, but people in conversation analysis and ethnomethodology did not like it that much, because it was not technical enough for their concerns. Our book was wider than just a textual analysis: it included participant observation of sellers. I returned to that work more recently, partly because we have done a new edition of the book.[4] Colin Clark, my coauthor, had this really clever idea. The book was out of print. There are probably many out-of-print sociology books and with online publishing you can start to reproduce them. He formed his own publishing company, called Sociografica, and our book was the first book of this new publishing house. It has a digital version and it is also print-on-demand, so it is very cheap to buy.

For this new edition we wrote two completely new chapters. One new chapter deals with the changes in the world of pitchers, because they have almost disappeared in Britain. This has to do with the decline of street shopping, and with the fact that supermarkets, online, and television shopping have grown. In another chapter we trace the actors we dealt with: some of them have moved to American infomercials. Infomercials have grown incredibly, and several of the pitchers that we studied have moved in that industry. We reinterviewed them, focusing on the transfer of their skills from the street markets to major media companies. We describe what made them leave the markets in England to work for television in America,

4. Clark and Pinch 2014.

and the peculiar skills in their biography that enable them to do that. They are supremely confident that they can sell anything. Moving to America seems impossible for them, but they think that the bigger the adversity—or, as they say, "the bigger the lie"—the more chances they have "of getting away with it."

So you have a narrative about their life course. That is one part of it. The other part of it is about how they translate their technique of selling, that they developed in street markets, to a different medium. The materiality is different. You do not have a live audience. You have an audience that is mediated by television: they are at home. So the problem is how do you stimulate demand in such an audience. For example, one of the strategies in face-to-face selling is to manufacture scarcity. They say things like "I can only do this price for the first five, ten people maybe. I can't do it for the eleventh person, you must buy now." On television that becomes a limited period of time. They may have this clock ticking down. Another example: In face-to-face selling, it is important for people to see other people buying. It compels them. There is nothing better than to see someone else buying, if you have doubts. So on television they show the phone bank, and they have the sound of the telephone calls coming in. In this way, people know that other people are calling. It's also about sound, of course!

ST This attention to the translation of skills is actually very interesting.

TP The attention to skills and knowledge is a very relevant part of my work on selling and materiality. I am currently interested in the notion of "field knowledge" or "territorial knowledge" in selling. This notion comes from studying with a graduate student, Ling-Fei Lin, how laptops are manufactured. Her PhD dissertation has been on laptop manufacturers in Taiwan and China.[5] As you probably know, most of the laptops and tablets today are manufactured by Taiwanese companies operating in Taiwan and in China. Her question is very interesting: What are the skills that these Taiwanese companies have? Like selling, technological innovation in production has always been underplayed. There is a global politics of race behind this: even if many innovations are coming from the East, from the Far East, the trendy innovations are only seen as the ones introduced by Intel or Apple: the software, or the architecture of the stuff inside the computer.

5. See Lin 2015.

Innovations in production are underplayed, even though many of them are crucial. If we have these devices cheaper it is partly because we have production innovation.

So the question concerns the skills these people have, and she came up with a whole new notion of a form knowledge which she called "field knowledge." It is a kind of territorial knowledge that enables these people to switch laptop production between places, first to east China and now to the western parts of China. In my opinion this is a very interesting form of knowledge. It is not tacit knowledge, it is territorial knowledge. I realized in talking with her—we plan to write a paper on this—that this knowledge is similar to the one that some sellers have. There is a kind of seller called a "field seller." They go out in their car and they have a knowledge of a certain territory, and the skills they have are limited to that geographical region.

In another paper I am coauthoring with Paula Jarzabkowski and Steve Hilgartner, we are finding that insurance people have this knowledge as well. They become experts in a territory. There is a huge amount of money to be made in this skill, knowing a territory, because you know what the risks are: the way houses are constructed or if trees are likely to fall on houses. This form of knowledge deals with a complex material world, and is limited to a certain territory. In another geographical region, insurance risks can be completely different, there might be people falling on ice and breaking their necks—something completely different. It is very interesting. It has to do with materiality and knowledge.

ST   How does this form of knowledge play a role in laptop production?

TP   In laptop production people often simplistically say that the reason to do laptops in China is because labor cost is cheaper. But Ling-Fei Lin has shown that there is a much more complex story than just labor cost. Labor cost is certainly part of it, but you have to find the right people to employ, this is clearly another part of it. What she argues is that the production engineers that are responsible for moving the factory have a regional knowledge, not only of labor but also of sources of supplies. She shows that when they move, first of all they go around, they visit the factories in a certain region. They started off in Japan, because a lot of laptop production was done in Japan. They used to go around, from company to company, and observe how the Japanese were doing this factory production. They get this knowledge from this particular region and they transplant it into another region. This is something that is very interesting: a territorial knowledge, and skills to transplant it elsewhere. But this idea still needs to be worked out: this knowledge is cutting edge—it is not like the social construction of technology that has been existing for thirty years! We are working on it right now, so it is not pinned down yet.

ST   It is very interesting, indeed. This "territorial knowledge" reminds me of Italian "industrial districts," a notion that dates back to Marshall.[6] Italy has got a tradition of small and medium enterprises that are very connected to their territories. In the late 1980s and early 1990s, Italian industrial districts have been widely studied as an interesting form of post-Fordist flexible production. The Italian literature on the topic is very large, but there is a lot also in English: Ash Amin[7] has written about this topic in economic geography, pointing out that there are relevant forms of tacit knowledge at play. Also Robert Putnam[8] or Charles Sabel and Jonathan Zeitlin[9] have published on the topic. If I have understood well your notion of "territorial knowledge," it plays a key role also in Italian districts, because in flexible production you have to know the resources of the territory, and you have to know where the expertise is. You have to know well other companies and the labor force. Moreover, the local dimension plays a key role in terms of personal relationships, of trust, and of formal or informal protocols of behavior. All these things are connected to a territory.

TP   Yes, this resonates with what we are doing, and it is very interesting. There is a new interest in production as well, not only in selling. Suddenly there is this interest in making. Production is becoming a trendy topic again, and so revisiting this older literature on post-Fordism is interesting, especially when it is about this kind of knowledge. Anyhow, I have also a new paper coming out that connects my interest in selling with sound studies.[10] You know: part of the materiality of selling is sound. The paper is called "The Sound of Economic Exchange" and is coming out in a reader, the new edition of *The Auditory Culture Reader* by Michael Bull and Les Back.[11] It starts with the opening and closing bells of the stock markets as

6. Marshall 1919.
7. Amin 2000.
8. Putnam 1993.
9. Sabel 2002, Sabel and Zeitlin 1985.
10. Pinch 2015b.
11. Bull and Back 2015.

iconic sounds, and it draws on the work of Donald MacKenzie[12] who has been studying financial markets. In his work he describes the importance of sound, like the calls of the brokers when they sell or buy in the London market. The sound is important: brokers have to develop a special ability to listen to conversations across the room. This is another important sonic skill, referred to as "brokers' ear."

ST This reminds me of something I was told during the interviews I did for one of my first papers,[13] on the language of the stock market's pits, in Milan. I remember I had these long interviews with some brokers that were active during the transition to electronic trading, in the early 1990s. One of the very interesting things they told me was that they could get an "overview" of what was happening in the market just by listening with a part of their attention to the background noise of the pits. A sudden change in the acoustic environment made them immediately understand that something was happening, and sometimes they could even get what was happening from sound. After the transition to electronic trading, they translated this skill to the new material arrangement of trading, adding sounds to the software they were using. They associated a specific sound to a specific event. They had many different sounds, so they could somehow replicate that skill of "overviewing" the situation thanks to background noise. Unfortunately, at that time, I was interested in body language, so I didn't ask them to record the sounds of their trading software, but they talk about this topic in the interviews, and I have mentioned it in the paper.

TP That is interesting. Is it in English?

ST Unfortunately not; it came out in an Italian edited book on the language of the pits.

TP You have to know that in this new paper I make exactly the same example you mentioned in your beautiful paper. I have talked to someone else that has done a similar study, Alex Preda, and he alleged exactly the same thing: that people were sonifying their machines. There is also this ethnography by Catlin Zaloom,[14] that followed the transition to electronic trading in Chicago and London. She has great interviews on the relevance of sound. She describes how people would practice their voice in front of

12. MacKenzie 2008.
13. Tosoni 2003.
14. Zaloom 2006.

the mirror, so they could shout in the right way. The book is called *Out of the Pits*, and it is a great ethnography. Sound is a very important part of the materiality of selling. My work on selling fits also into the sound studies work I am doing.

ST It seems that from discussing selling we have smoothly moved to sound studies, so I would give a closer look to your recent work in the field.

4.2 The Materiality of Sound

TP As I said, my interest in selling and materiality links to sound because sound is part of the materiality of selling.

ST Could we say that your interest in sound studies dates back to your work on the Moog synthesizer?

TP In hindsight, doing sound studies after my work on the Moog synthesizer seems the obvious thing to do, but you have to know that actually it took me ages to do that. A long time ago—I think it may have been 1996—I organized two sessions at the 4S/EASST Annual Meeting in Bielefeld called "STS faces the music." My intention was to do a special issue of *Social Studies of Science* about this. I talked with David Edge, the editor of the journal, about it but somehow we couldn't get it to work. In hindsight, I realize that the bigger subject was not music: it was sound, sound studies. I became interested more and more in sound, and now I realize that one sense, the visual sense, has dominated a lot of work on science and technology studies, and in sociology more generally. This is a view shared by many people now: there are other senses to "look at." I have become interested in how the different senses work together. If you start looking at science, technology, and medicine, you realize that sound plays a really important role in these fields. But it took me ages to get from thinking about the synthesizer study in terms of music to the bigger issue of sound.

Collaborating with Karin Bijsterveld at the University of Maastricht has been important for this: in particular, the paper we published in *Technology and Culture* on the introduction of new technologies in music as breaching experiments.[15] There, my interests were starting to broaden out, because

---

15. Pinch and Bijsterveld 2003, 537:

The approach we adopt in this article is to see the introduction of new technologies into music as a set of "breaching experiments." Such experiments, first introduced into sociology by Harold

we were looking at breaches in normative practices. We had three examples: the synthesizer, mechanical instruments like the player piano,[16] and
one of Karin's case studies, the *intonarumori*[17] of the 1910s by the futurist
Luigi Russolo. In collaborating with Karin I started to realize that there was
a bigger field. So we finally did a special issue for *Social Studies of Science*.[18]
The one on music didn't work, but this one was on sound and it did work.
It had papers on new technologies and instruments,[19] on practices and
on language in recording studios,[20] and on tacit knowledge in recording
studios.[21]

This last piece by Susan Schmidt Horning was important because I
remember, writing the introduction, that I thought: "this is about sound."
So it was important to define sound studies. I remember sitting in my office
and coming up with what I thought was a great definition of sound studies: "sound studies is an emerging interdisciplinary area that studies the
material production and consumption of music, sound, noise, and silence,

---

Garfinkel, serve as probes into how everyday social order is maintained. ... We want to treat the
introduction of new machines as prima facie cases of breaches in musical culture. Such interjec-
tions provide an opportunity to rehearse arguments about what counts as part of music and art
and, conversely, what is appropriately delegated to machines; breaches of convention reveal
underlying norms and values. ... Reactions to new technologies could provide a fertile research
site for investigating how technologies in general are embedded within conventional normative
frameworks.

16. Introduced at the turn of the twentieth century, "player pianos, such as the
Pianola and the Welte-Mignon, produced their music mechanically through a set of
instructions stored on a perforated music roll. Their commercial success depended
upon not only the standardization of music rolls but also important changes in cultural values" (Pinch and Bijsterveld 2003, 540).

17. "According to the musicologist Barclay Brown, who describes Russolo's instruments as some of the very first synthesizers, the noise-generating parts of these
instruments were powered by electric motors, handles, or hand bellows, and each
sound-generating device produced a characteristic noise. Furthermore, each instrument had a mechanism to amplify sound and control pitch—a sort of chemically
prepared drum skin on a frame with a vibrating wire attached to the center of the
skin, creating a specific pitch. Some of these noise-producing mechanisms were
rotating disks with differing surfaces that caused the wire to vibrate" (Pinch and
Bijsterveld 2003, 544–545).

18. See Pinch and Bijsterveld 2004.

19. Bijsterveld and Schulp 2004.

20. Porcello 2004.

21. Horning 2004.

and how these have changed throughout history and within different societies."[22] It is interesting how you define a field: it was much broader than music. Now, when people want a definition of sound studies, that is the one they often go to. The music department of the Harvard University has this quote from me and Karin Bijsterveld on the website of their important series on sound called *Hearing Modernity*.[23] There is a funny story about this: they attributed it to Murray Schafer! And they dated it back to when Schafer developed the concept of soundscapes. So they had Murray Schafer having defined the field of sound studies like in the late 1960s. Karin had noticed this before me. I remember I sent them an email saying "Hey! Actually it is Pinch and Bijsterveld and it's not from the 1960s or the 1970s!" They were rather embarrassed and they changed it to us.[24] I have seen that definition in several places now. In the *Oxford Handbook of Sound Studies*[25] that Karin and I edited for Oxford University Press, we used again that same definition. I cannot improve it: it is a great all-encompassing definition of sound studies.

ST What do we find in this definition of sound studies?

TP First of all the definition includes music, sound, noise, and silence. So we are away from just music, we are into the bigger domain of sound and of absence of sound, which is silence, and their material production. You may think that silence is simply no sound, but if you think about it, silence has to be produced as well, because silence is never absolute silence. Actually, you have accomplishing of silence—you have to think about it in terms of accomplishing. For example, you can think about the accomplishing of silence in football stadiums. It is incredibly moving: you have it to commemorate some tragic event, such as the Hillsborough football stadium disaster[26] where many Liverpool supporters lost their lives. You have a minute of silence: how is a minute of silence accomplished? It is never complete

---

22. Pinch and Bijsterveld 2004, 636.
23. See http://hearingmodernity.org/ (last accessed Apr. 11, 2016).
24. Although now correctly attributed to Pinch and Bijsterveld (2004), the definition is still dated 1977. See http://hearingmodernity.org/sawyer-seminar/ (last accessed Apr. 11, 2016).
25. Pinch and Bijsterveld 2012.
26. The Hillsborough Stadium disaster (Sheffield, England, Apr. 15, 1989), where 96 people lost their lives and 766 got injured, is considered the worst disaster in English sports history.

silence. It is a strange ambient sound. It goes from this roar to silence, and it is incredibly moving because a contrast is accomplished. Even in movies, when there is silence, it is not complete silence. There is a special technique for rendering silence in movies: there is actually a level of white noise in there. So how silence is accomplished is very interesting.

Second point, the definition refers both to production and consumption, to production and to listening experiences in different cultures.

Finally there is a stress on materiality, on how the materiality and the culture of sound production and consumption are entangled together. The *Oxford Handbook of Sound Studies* has a lot on the role of materiality and technology. Actually, it is what science and technology studies are good at dealing with. Other approaches to sound studies tend to black-box those aspects. Coming from STS, we are particularly good at thinking about the materiality of sound and the technology of sound.

ST What is the specificity of a STS-inspired take on sound?

TP Our main argument in making that book is that sound is becoming much more materialized, more thing-like. For me, one of the most important realizations I have come to is that there is a difference in the language and skills that we have to deal with sound and sonic experiences, and the visual. Even when we edited that book I don't think I had fully seen the significance of this. This difference is simply that the means of taking the sonic from its original source of production and transporting it around the world have come fairly recently, with the development in the 1870s of the phonograph: that's when it starts. For the visual, with painting, it has been much longer. Painting could be transported away from the material place where it was produced, it could be looked at in another place and it could be copied way before you had sonic copying. The visual has been copied and reproduced much earlier, with the printing press, and I think that as a result we are much better able to describe it: there is a more sophisticated language for dealing with the visual than with the sonic. It sounds like a crude argument, but there is something in it. There is much less language for discussing sound than there is for the visual.

This makes the whole language of sound a very interesting topic. Why for example do we have "fat" sound versus "thin" sound? We don't really know. It is strange because in the recording studio people value fat sounds, it is one of the few contexts in which fat is good and thin is bad. Usually in most other contexts fat is bad and thin is good. How does language get attached to sound is an interesting topic, and I see it as a part of a more general interest in how sound stabilizes.

Recently I started to think back to *Analog Days*, the book on the synthesizers you were mentioning, which I wrote with Frank Trocco. We had a quote there: "sound is the biggest silence in this book."[27] Actually it was true: I realize we didn't really have that much to say about sound. It was mainly about how the technology stabilizes, how it is sold, and how it spreads. We had a whole account of that, but we didn't say so much about how sound spreads. My recent work has been a return to the synthesizer case study to think about what was missing there: sound, how sound emerges, and how does it stabilize. In my first talks on this topic I started to deal with sounds that are somehow visualized, because I thought it was easier to deal with them. So, I looked at the use of the synthesizer in *Star Wars*, for making spacecraft sounds, like the sound of the Millennium Falcon, and many other iconic sounds. This was the work of the sound designer Ben Burtt. The term "sound designer" was first developed after *Star Wars*, and sound is incredibly important for Lucasfilm and for Pixar. There is an interesting article on the key role played by sound in the success of the Pixar company: this argument was made by a film studies scholar called William Whittington, who has a piece[28] in our *Oxford Handbook of Sound Studies*. I talked about those iconic sounds because there you have a visual reference as well, but now I am doing the much harder part of the project: I am talking about sounds that do not have a visual reference. How do they stabilize?

ST I can see how this development stems from SCOT and from your previous work on technology: as technological artifacts, sound also has issues of stabilization.

TP Exactly, and the article where I think so far I have best developed my thinking about how sound stabilizes is the one that I have published in a book edited by Ariane Antal Berthoin, Michael Hutter, and David

---

27. "If we have told the story well, we will have brought to life the part played by one machine in shaping our culture, and the part played by our culture in shaping this one machine. The paradox is that our story is about sound—and words alone can never express what it was like to hear for the first time the beat of a pair of oscillators through a big sound system, in a vast arena, in the early morning, after the rain. Sound is the biggest silence in our book" (Pinch and Trocco 2002, 11).
28. Whittington 2012.

Stark.[29] It is a collection on valuation that has recently come out for the Oxford University Press.

There is a great piece by Antoine Hennion[30] about wine tasting in that book, on how people are evaluating wine. To evaluate something you have to think about the senses. Wine is valued by people who taste it and smell it, to some extent. That's how you tell if it is a good wine or a bad wine. What is very, very interesting about Hennion's paper is that he talks about the capacity of wine to surprise: it must have a capacity to surprise you. That is a really good point: it points again to the materiality of wine that can have that capability. So I started thinking in detail about the difference between wine tasting and sound, because sound as well has a capacity to surprise. There is a great quote from Robert Moog on the first sounds of the synthesizer.[31] He describes how people are surprised by that sound, because it is something new: something new is added in the world of sound when you have electronic sound. They are hearing that sound for the first time. Why is it surprising? How do you start to describe that? How do you develop a language in sound studies or science studies to describe what those sounds are?

What you have there is an interesting combination of the place where the sound is produced, the technology that is used to produce the sound, how the sound is heard and the means to reproduce that sound so it can be heard again and again by many people. A little bit of it is actually describing the sound, but it turns out that the description of the sound is not the most important thing, because you can hear it. I can play to you the bass sound that Simon and Garfunkel had in their album *Bookends*,[32] which is also the "same" bass sound—same in inverted commas—that Moog produced for them in the studio. I don't have to have a language to describe it, as I can play that sound, you can experience it.

29. Berthoin Antal, Hutter, and Stark 2015.
30. Hennion 2015.
31. "He [Deutsch] took this and he went down in the basement where we had a little table set up and he started putting music together … I still remember, the door was open, we didn't have air conditioning or anything like that, it was late spring and people would walk by, you know, if they would hear something, they would stand there, they'd listen and they'd shake their heads. You know they'd listen again— what is this weird shit coming out of the basement?" (Interview with Robert Moog, June 5, 1996, quoted in Pinch 2015a).
32. Simon and Garfunkel, *Bookends* (Columbia, 1968).

So I started to think about what are the key sounds of the synthesizer that start to stabilize, or to have value, in a musical culture. How do you find it? You talk to Moog, or with the session musicians using the synthesizer for the first time. There is a great account by Ray Manzarek[33] of the Doors about how the Moog synthesizer is used on *Strange Days*,[34] the second album of the band, and the first album where they use the synthesizer. The synthesizer player was Robert Moog's West Coast salesman, Paul Beaver. He is in the studio, producing the sound and Manzarek describes the process. The sound is not yet stabilized, that is the interesting thing. So Jim Morrison, the singer of the Doors, is hearing a sound he wants to use on the record and he can only describe it as if he has been tripping out on acid. He describes it in an incredible way: he says it is a crystalline sound, like the "sound of broken glass falling from the void into creation." So he is using these words to grasp towards the sound, and Paul Beaver translates that: "Oh you mean the sound I made about three back?" And the producer says: "Can you go back to that sound?" He tries but he realizes he can't go back to it, because it is not yet stabilized. It is a beautiful example. They are hearing the sound they want but it is not yet stabilized.

In that paper I am trying to describe how you get from that to a stabilized sound where the artists and the producers can say "that's the sound we want": they can reproduce it, and listeners can recognize it. How does that process happen? To answer I take some iconic sounds. Two sounds: I take this bass sound, which Simon and Garfunkel used for the very first time, and that later became the sound of hip hop. It is a really recognizable moog-based sound that is used in electronic dance music. So I take that as a first iconic sound, and I take a yawling sound from progressive rock that Keith Emerson and Rick Wakeman can make using the Minimoog synthesizer. It is a piercing monophonic sound with the property of glissando, it slides. To present this paper, I found a YouTube video[35] of the rock band Heart. In their famous song *Magic Man* there is a little Minimoog solo. In the video you can actually see their guitarist play the Minimoog, and you can actually see the technique he uses. Here there is a performative aspect of a particular genre of rock, and it is important for how sounds get value.

33. Manzarek 1998, quoted in Pinch 2015a.
34. The Doors, *Strange Days* (Elektra, 1967).
35. See https://www.youtube.com/watch?v=Ps7tVvQHLyo (last accessed Apr. 11, 2016).

There is the dress, the ambiance, the guy: he puts the guitar on his back and stretches out to the keyboard making this piercing monophonic sound that takes over from the guitarist. It is an incredible sound: you can listen to the sound and you can see how it is produced, and technically describe it. This is a key sound of prog rock from that time.

I am quite happy with this work because it is really hard to describe how sound stabilizes. You have to find which are the iconic sounds. I got two iconic sounds from a particular piece of technology and I worked my way back in analyzing and describing how these sounds became valuable, and then these sounds are reproduced and other people say they are what they want of the Minimoog. They want that sort of sound that Emerson and Wakeman can do. There are many other sounds, but it is just hard to do two of them, trust me.

ST   Talking about stabilization, how stable is the field of sound studies at this point? Is it possible to identify distinct schools or subfields of research, or is the field still more chaotic than that?

TP   I guess it is still more chaotic. It is like in the early days in SSK, before the different schools crystalized: there is a lot of agreement in the field, and there is even a bad old philosophical approach to sound that nobody wants. But there is also a "good" philosophical approach to sound, developed by Don Ihde in his wonderful little book, *Listening and Voice*.[36] That book can be seen in hindsight as one of the first books in sound studies. Basically, there is an enormous amount of agreement, even if you are starting to see some of the differences of the approaches. There is this guy called Brandon LaBelle[37] who has done ethnography of sound, who is based in Berlin. There is Michael Bull[38] in Brighton who does mobile sound—Walkmans, iPods, and so on. There is a guy called Steve Goodman[39] who has done this book on sonic warfare. There is my STS colleague Karin Bijsterveld[40] and

36. Don Ihde is Distinguished Professor of Philosophy at the State University of New York at Stony Brook, and a pioneer and leading scholar in technoscience and post-phenomenology. *Listening and Voice* (1976) is a groundbreaking classic in the phenomenological study of sound.

37. See LaBelle 2010.

38. See Bull 2000, 2008.

39. Goodman 2010.

40. See Bijsterveld 2008; Pinch and Bijsterveld 2012; Bijsterveld, Cleophus, Krebs, and Mom 2014.

her group in Maastricht who work on sonic skills. Emily Thompson, who comes out of history of technology, has a wonderful early book on sound, *The Soundscape of Modernity*.[41] Jonathan Sterne has been enormously influential with his excellent book, *The Audible Past*,[42] which shows how sound becomes a commodity, and with his new book on the MP3 format.[43]

I endorsed that book, by the way, but I am now rereading it and I can see a few differences between what Sterne is doing and what I am doing. The approach that Sterne takes is slightly different. Although he is into describing the details of the technology, he goes into what I call a "media history" approach. I am slowly realizing that some of these scholars, such as Lisa Gitelman[44] at NYU, are interested in what you might call "media architecture." They are doing history, but they are depicting a kind of backwards history of how a technology gets from A to B. They seem to start with the B, and they tell how we get there from A. Mara Mills[45] at NYU, who has done incredible historical studies of sonic technologies and has opened up a whole new area of disability studies, seems to take this approach. So Sterne takes for instance MP3, and he traces the history back to Bell Telephone Company, to describe how this algorithm has evolved.

That is fascinating stuff but it is slightly different from a STS approach. In STS we are doing a history that is much more contextual. We are interested in all the paths that were not taken. We would start back in the 1930s and look at what was happening at Bell and see all the possibilities and the paths not taken. They start with a modern technology, what they call a media architecture, and they work back. They open the black box to some extent, but I find it a little dissatisfactory as a method of historical analysis. But Jonathan Sterne's work on tracing what he calls "audile techniques" through the stethoscope, telegraph, and telephone, and how this leads to the commodification of sound, is very compelling. Sterne has really been hugely influential in getting sound studies on the map. The approach that we are developing, Karin Bijsterveld and me, with our students, is more of a traditional STS approach, where we go to a particular context and we explore the paths that were taken and that were not taken.

41. See Thompson 2002.
42. Sterne 2003.
43. Sterne 2012.
44. See Gitelman 2006.
45. See Mills 2012.

That's why I say it is a more contextualized history. It is closer to the SCOT approach. I don't know if this makes any sense to you, because you are a media scholar.

ST It makes sense, but actually I was wondering why you call that a "media history" approach. I see these as two different methodological approaches to history, and not as related to specific research objects like media. I acknowledge that media scholars tend to work in a more linear and genealogic way, but I think it would be great to have more "contextual" historical accounts also for media.

TP Yes, I guess you are right.

ST There is another question I would like to ask you about the field. So far, all the approaches and the works we have discussed in the field of sound studies deal with sound as their main research object. I'd like to ask you how developed the attempts to address sound are as part of broader social processes. I mean attempts to integrate into a broader analysis of social processes an attention to their sonic aspects, where sound is relevant, but it does not necessarily represent the main research object. Personally, I came to realize how relevant it is to try to integrate our sociological approaches with an attention to sound some years ago, when I participated in a rather big research project on daily life in different Italian neighborhoods. As a media scholar, I was asked to work on representations and perceptions of insecurity,[46] and sound, what people could hear, played a very important role, even if not an exclusive one. I remember people in Genoa describing the sound of cars running wildly in the streets at night or, in a more dramatic way, some families in Palermo describing the sound of violence coming from the apartments of people living in the same building. Those noises were very important for the perceptions people had of their context and, as a consequence, for how they lived their everyday lives there. Moreover, multisensoriality is becoming a big issue now in media studies, especially for the analysis of media practices in urban space: that is the topic I am working on.

TP You are right, that needs to be done, and indeed it has been done by historians such as Mark Smith.[47] Well, I am trying: if you think about the work on selling, selling is a social process. I am trying to bring back the

46. See Tosoni 2007.
47. Smith 2012.

acoustic dimension into the economic exchange. Of course we could do a lot more, as you show in your examples. For instance, I just read a paper on sound and class. Someone was analyzing sound complaints in an area of a city in terms of class differences in neighborhoods. There is an association between noise and working class, just to make it crudely. What is a tolerable, or even desirable sound and what is not is related to class. So you have bourgeoisie sounds, like someone playing a bit of Beethoven that can be heard out of a window. That's an acceptable sound, but a guy with a ghetto blaster, an African American group of people standing on the street or doing rap, is seen as a bad sound. Sound of violence is a bad sound. So you are right, people need to do that sort of work. I am starting to see that sort of work and I am trying to do it in the context of selling because I have some expertise in there, but that needs to be done. Sound needs to be reembedded from just this specialized area into the wider sociology. In my vision you should do that. And by the way, it's not just sound; there are also the other senses.

ST Yeah, multisensoriality.

TP The thing is about senses. Taste is an important part of it. As I have already mentioned, Hennion has this work on wine tasting, but there are also other people. Latour has this PhD student, Thomas Vangeebergen, who is looking at meat pies in a working-class factory, because they have a science of taste there. If you want to make a meat pie cheaper, it still has got to taste like a proper meat pie. So we are not talking about refined, high quality wine here, but still it has got to have a certain taste: you have to find out what you can substitute to make the production cheaper. It is interesting to have experts on taste in these meat pie factories.

ST What about the methods and the methodology employed to do research on sound? What are the main challenges and difficulties? For example, I know that in the *Hard Sell* you used video recordings. How do the analyses of source materials relate with the interviews, and with social actors' representations?

TP This has to do with the difference in recording the sonic and the visual we have discussed. When it is possible, it is great to work with both source materials and interviews. Revisiting the *Hard Sell* to work on the sound of selling, I went back to the videos we recorded, because the sonic is there. When we made the videos, technically, the hardest thing to do was to get good sound, because in the street markets it is very easy to make good quality video but the sonic is very hard to do. We are pretty proud of it. In that paper on the sound of economic exchange[48] there is a part of the routine that is called "twilling the edge." The "edge" is what the pitchers call the crowd gathered at the stall. Once one sale is made, the pitcher spins it out like a twirling top, trying to draw more people in the sale, and sound is actually crucial in this. We have a beautiful example of that from Harringay market in North London. The pitcher has got a microphone and he has got a crew. They are selling dolls, twin dolls, and at each sale the crew shouts and this is building up to a cacophony of sound, as they are twirling the edge. They shout "the pair!"—since they are twin dolls—"another one, another one," "again!," "over there!" and they get louder and louder. We have the sound recording: you can hear that cacophony. It is a sound I can play back when I am giving academic presentations, and you can hear that wall of sound build up. That sound is building up the excitement, and you can still hear that sound, even if it still leaves the problem of whether the sound that is heard today in a recording is heard as it was heard in the 1980s.

Yet, sometimes you don't have these source materials, and you have to rely just on interviews. For example, the noise of the pitchers and how it excites people ties in beautifully with the earlier studies of Mark Smith, the sound historian who wrote about the soundscape of early industrialization in Massachusetts. He compared the soundscape of the industrial revolution in America with the soundscape of the South, which is a much quieter soundscape. There are no recordings, but the interesting thing he found is that the women going to work in what we would think of today as incredibly loud factories, actually found this sound exciting. It is the sound of modernity, of progress, of industrialization, compared with the rural sound. He went to their diaries, and rather than complaining about the sound of these factories, those women were actually welcoming that sound. He contrasts that with the sound of South, of slavery and plantations.

He had only access to the diaries, but working on the synthesizer I have the actual sounds, I can listen to these sounds. For me it is methodologically really important: I have to listen to the sound. I like to playback the sound when I give a talk. That's why for me it is really important

48. Pinch 2015b.

to have good sonic equipment there, and there is always the issue: will it sound the same? Because it never will, you know. Listening to Wendy Carlos in *Switched-On Bach*,[49] that Robert Moog played at the Audio Engineering Society[50] with very high-quality speakers and the original master tape, was a very different experience from listening via a little MP3 file at an academic talk, with a couple of computer speakers. It is a very different sonic experience—the whole context is different—but I still think that you should make an attempt to play back and listen, and at the same time point out to the audience the problems of faithfully reproducing sounds from a different period and location.

ST Yet the written article or book is still the main communication format used within academia. This may not be a problem if you are dealing with diaries or interviews, where sound is already mediated by social actors' representations, but in other cases you have a problem of translation. Moreover, you have already pointed out the problem of lacking a refined vocabulary to talk about sound, if "talking about sound" is the right methodological thing to do.

TP Yeah. The only thing that is going on now is that some of these articles I referred to are published with audio files. There is a new journal of sound studies where I published an article, the *Journal of Sonic Studies*, where you can embed sound files in the article: I have done that. The reader of the article can hear the sound file as well. We wanted to do that also with the *Oxford Handbook of Sound Studies*: we insisted on having a good companion website with sound files. The trouble with all that stuff is that some people give you the sound files on YouTube and then the YouTube videos are down for various reasons. It is very frustrating. If it is a sound that you can put yourself in a journal that's fine, but if it depends upon a commercial platform like YouTube, then you've got also new problems: annoying ads and so on.

ST Under an institutional point of view, how is the field of sound studies going? Is it taking shape? Are there courses in universities?

TP It is growing rapidly. Here is a funny thing: I am giving more and more talks in music departments. Music departments have become very

49. Wendy Carlos, *Switched-On Bach* (Columbia, 1968).
50. Robert Moog played an excerpt of *Switched-on Bach* before its release after a talk given at the Audio Engineering Society conference in 1968.

interested in sound studies, partly because they are realizing there is a materiality in musical instruments, which is leading them to read books like *Analog Days* in a new way. We have two people at Cornell, in our music department, who are very interested in science and technology studies, Ben Piekut and Roger Mosely. I actually think that sound studies in music departments is going to take off. That is a natural place for it. Another part of this is simply that students who are coming to music departments these days are used to the digital tools of recording, and messing with sound that way makes them realize that the bigger topic might be sound. "Why am I making such a fetish of one instrument?" I think that the technology and people's skills in using the technology are kind of opening the black box of music as a subject much more. Some are very excited by this. Then there is this new journal on sound, the *Journal of Sonic Studies* I mentioned that seems to be getting some traction. Michael Bull is editing a new series of books on sound and another new journal called *Sound Studies*. I think sound studies as a field is really starting to take off.

ST   What about access to funds, both public and private? Who are the main actors interested in the field, if there are any?

TP   Well, there is a huge interest in sound, but I have not been approached by any companies yet. Well, that's not quite true: I often advise students with startups, and some of these are developing new sonic technologies such as Thinkplay,[51] which has a software for turning your guitar into a synthesizer or multimedia controller.

Here is another interesting fact: a car company like Porsche employs more audio engineers than any other sort of engineer, and that tells you about the importance of sound in modern automobile construction. The door must sound right, the seat must sound right, when you sit on a Porsche seat. Also the engine must sound right: a Porsche engine must sound like a Porsche engine. Karin Bijsterveld has a new book coming out about that.[52] What people look for in sound is different between different countries as well, it turns out. So Porsche have designed the engines to be slightly different for different countries' consumption. There are also people working on sonic warnings, because for obvious reasons they are better than visual warnings: visual distracts you. They are studying things like the best voice

51. Http://thinkplayaudio.com/ (last accessed: Apr. 11, 2016).
52. Bijsterveld forthcoming.

for a warning in a BMW. They found out things like a woman's voice is more effective than a man's voice. They also study what is the best phrasing. If you are driving too fast you should not have a warning that says "Stop!" or "Slow down." It should just be a reminder of your speed: it seems it is much more effective. "You're now traveling at 80 km/h," something like that.

Commercial companies are interested in these studies, but they are not funding studies like mine, or like the ones we have talked about. Sound studies do not have an enormous commercial impact. Very little private money is funding fundamental investigations into the nature of sound and its link to culture, so I don't think sound studies are going to be huge in that way. It is more from the academy that the interest sparks.

4.3   New Developments in STS: The Ontological Turn

ST   As a way to conclude this interview, I would ask you what is currently going on in the field of STS. What are the most important debates at the moment?

TP   Well, there is an interesting approach developed in Britain and the Netherlands by John Law and Annemarie Mol,[53] coming out of actor-network theory. Their approach is a turn toward looking at ontology. This turn comes from the idea that material agency should be given more weight. It is generally known in philosophy as posthumanism. So it is a turn toward a posthumanist approach, and Steve Woolgar and Javier Lezaun edited this special issue of *Social Studies of Science* on ontology[54] on it. There is a lot of interest in ontology. It stems also from Andy Pickering's work, *The Mangle of Practice*.[55]

I am quite skeptical about this turn, because all the approaches that turn to ontology suffer from the same problem: that you cannot distinguish ontology from epistemology. To say what is out there in the world always involves categorizing what is in the world and making those sorts of distinctions. You inevitably end up with representations in epistemology—you

---

53. See Mol 2002; Law and Mol 2008; Law and Lien 2013.
54. "A Turn to Ontology in Science and Technology Studies?" *Social Studies of Science*, special issue, 43 (3) 2013, edited by Steve Woolgar and Javier Lezaun.
55. Pickering 1995. See also Pickering and Guzik 2008.

want to call it epistemology, classification, or whatever you want to call it. It is inescapable.

A few years ago I did an essay review[56] of a book by Karen Barad,[57] who has published a whole book on linking quantum mechanics to feminist approaches to understanding what she calls "agential realism."[58] That is a position she has worked out, but I am very suspicious of a thing like that. Actually we had kind of a fight over that. For me it is wrong to use science as an underpinning for epistemology, and to invest this ontology with some sort of capacity.

Admittedly Barad talks about intra-actions, which are entangled relational properties, but I often get the sense that the world of objects is given more independent agency than within traditional humanist approaches, such as SCOT.

There is something similar going on in the field of sound studies as well. Some people try to pose an ontology of vibrations. Somehow this would be primary to the representational level. An example could be Steve Goodman, who wrote about sonic warfare.[59] He says that I can point a sonic gun

56. Pinch 2011. For an answer to Pinch's criticisms, see Barad 2011.

57. See Barad 1998, 2007.

58. Barad 1998, 89:

Agential realism is an epistemological and ontological framework that extends Bohr's insights and takes as its central concerns the nature of materiality, the relationship between the material and the discursive, the nature of "nature" and of "culture" and the relationship between them, the nature of agency, and the effects of boundary, including the nature of exclusions that accompany boundary projects. Agential realism entails a reformulation of both of its terms—"agency" and "realism"—and provides an understanding of the role of human and nonhuman factors in the production of knowledge, thereby moving considerations of epistemic practices beyond the traditional realism versus social constructivism debates.

59. Goodman 2010, xiv–xv:

Sonic Warfare outlines the acoustic violence of vibration and the trembling of temperaments. It sketches a map of forces with each step, constructing concepts to investigate the deployment of sound systems in the modulation of affect. The argument is based on the contention that, to date, most theoretical discussions of the resonances of sound and music cultures with relations of power, in their amnesia of vibration, have a missing dimension. This missing dimension, and the ethico-aesthetic paradigm it beckons, will be termed the politics of frequency. In order to map this black hole, a specifically tuned transdisciplinary methodology is required that draws from philosophy, science, fiction, aesthetics, and popular culture against the backdrop of a creeping military urbanism. By constructing this method as a nonrepresentational ontology of vibrational force, and thus the rhythmic nexus of body, technology, and sonic process, some latent affective tendencies of contemporary urban cultures in the early-twenty-first century can be made manifest.

at you, and you don't have to go through the process of representation to realize it. It is the same thing if I had a laser, and you'd say, "Oh I'm seeing a bright light," but it would be still doing damage to you. No one is denying there is a physicality, a materiality to sound and to light. That is an important part of what is going on. To understand what has just happened there is a representation through physics. But if you claim that to do sound studies you need this kind of philosophy of vibrations, it seems to me that you are arbitrarily giving primacy to that representation. You are simply privileging vibrations. A vibration is something you understand from physics. It has got a path in time and space. If you are into the language of physics representation and knowledge, a vibration form is a harmonic. In other words, things can be hard to represent and the specialist language of the sciences is the best way of representing what they are. But I find it very strange when people pose this new ontology of vibrations as something you need to study sound, it immediately alerts me.

This does not mean that materiality does not matter. As I told you, I have found this paper by Antoine Hennion[60] on wine tasting very interesting. He has this slogan: "matter matters." For Hennion the capacity of a good wine to surprise is the key thing to understand that matter matters. This point is very interesting, but this is also a very tricky issue, because matter matters, but it doesn't matter in some sense. Here is my example: suppose you have an unsophisticated palate and you can hardly tell the difference between a Latour wine[61] and a plonk. We all know you have to learn how to taste wine. This is a skill. For a naïf taster some plonk by the supermarket and a Latour wine could be hardly distinguishable (I like to use that Latour wine for this example, since Hennion comes from Latour's school and actor-network theory!). We are not sure that an unsophisticated palate would be "surprised" by a Latour wine. So matter matters, but it does for a skilled user. For an unskilled user maybe it does not matter that much.

I could give another autobiographic example on this. My wife Christine is not a big fan of electronic music, and for someone who does not like electronic music, probably the sound of the Moog, or of the Buchla, and all these different electronic sounds I have been talking about in such a

60. Hennion 2015.
61. Château Latour is a French wine estate, producing three renowned (and expensive) red wines. It turns out Pinch was actually referring to the Latour bottling label owned by Bruno Latour's family—this wine is a Burgundy.

nuanced way, probably sound very similar. It is clear that with electronic sound you have a new sort of sound, a new type of experience, a new phenomenon, something new has happened in the world. I am prepared to accept that new things emerge in the world, and you can study this issue of emergence. If you want you can talk about material agency, but to surprise, a good wine still needs anyhow a skilled taster. So that is again a cultural issue.

Although we can say that there has been too much attention on humans, and we have not considered the materiality of the world enough, I think that this ontological turn is going too far from the human side of the equation, trying to talk about what is out there in the world. Because to know what is out there in the world, always involves humans. Humans are doing the knowing here. So I am not opposed to humanism: I'm some sort of a humanistic sociologist, ultimately.

ST Unfortunately I am not a great expert of wines, but I wonder if this capacity of a good wine to surprise, as a form of agency of its materiality, is uncontroversial, as I also wonder if there is a universal agreement on which wines are those that are able to surprise you. Because if there is no agreement, it seems to me that we could be once again in an experimenter's regress situation. How can you evaluate the expertise of a wine taster, and at the same time the quality of the wine he defines as "good"?

TP Yeah, you're right. I didn't think about it, but you are right, it is exactly like the experimenter's regress. Once you have the skilled practitioners and you know this is a good taste, then the skilled practitioner can train you to taste it in the same way, exactly like when you repeat an experiment in the sciences. He can teach you to recognize a "surprising" wine. But if somebody disagrees, if somebody says that one wine is fantastic, and another tastes it and says that it is not so good, actually you have got an experimenter's regress issue. You can imagine it happening with a new wine, and probably it actually happens. That would make for an interesting study. There must be disagreements in wine tasting. Hennion didn't look at that! He is always assuming they agree. But I am sure they disagree, and you are right, it would be interesting to look at that.

ST There is another key point in this renewed interest in ontology within the STS field that I would like to ask you about. In the introduction of the special issue of *Social Studies of Science* you mentioned, Steve Woolgar and Javier Lezaun explain how the main impetus of this turn is the attempt to

avoid what could be called "perspectivism,"[62] or the "description and qualification" of different perspectives on the same ontological object. In their
paper on the Atlantic salmon published in the same special issue, John Law
and Marianne Elisabeth Lien use the interesting concept of "enactment" to
underscore how different practices enact multiple and different "salmon"
that subsist in multiple and different worlds.[63] In this way, they circumvent
an approach to multiple perspectives, or representations, or points of view,
on the same ontological object, in their case a "general salmon." I would
like to ask you to clarify the differences between this approach and social
constructivism in terms, on the one hand, of preliminary assumptions and,
on the other hand, of theoretical and methodological implications. Among
other things, I am asking this because for some scholars, like Patrik Aspers,[64]
the two approaches do not actually differ that much.

TP   Well, first off, there are some things about the so-called "turn to ontology" that are worth noting. As in all claims to do something new, and
in so-called new turns in particular, the old and "bad" straight road must
be distinguished before being dissed and dismissed. This is typically done
by saying social construction is all about "representation" and "perspectivism" and does not deal with the practices whereby different objects get
performed or "enacted." I reject these assertions for the brand of constructivism I am associated with. I have always dealt with practices whether in
science, surgery, or music. From the earliest days, we have been studying
how objects are entangled with practices and language (meanings). Early
on we used the Wittgensteinian phrase "form of life" to capture this. The
beauty of Wittgenstein is he shows how ordinary language, objects, and
practices are built together.

62. "The most explicit impetus of new ontological investigations in science and
technology studies (STS) is the desire to avoid being caught up in the description
and qualification of 'perspectives.' It is an effort to circumvent epistemology and
its attendant language of representation in favour of an approach that addresses
itself more directly to the composition of the world" (Woolgar and Lezaun 2013,
321–322).
63. "We do without the assumption that there are salmon out there with a definite
form, in existence outside the practices in which they are being done. That is the first
move. And then, here's the second, it follows that since those practices aren't the
same, different and multiple salmon subsist in different and multiple worlds" (Law
and Lien 2013, 366).
64. See Aspers 2015. For an answer, see Woolgar and Lezaun 2015.

If you reread one of the first major studies I did with Harry Collins, *Frames of Meaning*,[65] we were interested in how it was that for one group of scientists, paranormal phenomena did not exist, but for another group they did. In today's language we could say we showed how one group of scientists enacted the paranormal phenomenon of spoon bending and another group failed to enact it. We did this in a fully symmetrical way with an ethnographic method we called "participant comprehension" by doing fieldwork amongst both groups. It is true that we did not use the word ontology much, and used more the language of "frames of meaning" and "incommensurable paradigms," but scientific practice in a way is the ultimate means of doing ontology—it establishes what objects there are, or there are not, in the world. We drew direct analogies with Azande witchcraft practice; whether magic exists or not is a similar question to which we posed for paranormal metal bending. So we have always been dealing with questions about what objects and forces there are or could be in the world.

In fact that study is a lot more radical than today's claim for an ontological turn. Part of the concern today seems to be alterity—what might have been or could be in the world and is excluded and how to get at that. Studying paranormal phenomena in a way is the ultimate case of this. I think in hindsight we were quite sophisticated about these questions. The sorts of entities that are enacted in the world are tied in with "forms of life" and we saw that it was very likely that in the future, as we lived less amongst the parapsychologists and shared fewer of their practices—in other words, as we lived in a different form of life—it would be hard to recapture a natural attitude where all sorts of forces, such as backward causation, were routine. In short, we dealt not only with the enactment of objects, but under what sorts of circumstances they were unlikely to be enacted.

Next I want to turn to the point about multiple objects versus multiple descriptions of the same object. This always seems to me to entail a confusion—perhaps something fishy?—because John Law and Marianne Lien, in claiming multiple salmon being enacted in their study, have to say they are multiples of some entity with some sort of similarity—namely some sort of "salmon similarity." As Wittgenstein points out, it all comes down to similarity and difference judgments, and this is inescapable for them as well. There is no ontology without epistemology and vice versa. In making

65. Collins and Pinch 1982.

a contrast between the multiplicity of their enacted salmon on the fish farm versus the unitary, generalized salmon found in biology text books, again they have to posit some sort of similarity in "salmonness" to make the contrast. Some sorts of similarity judgments are inescapable.

Rather than frame the issue as being about multiple descriptions of the same object versus multiple objects, I would rather frame it in terms of why and when and under what circumstances do we get stability or multiplicity, and what sorts of objects or interpretations of objects might get excluded? For instance, on the salmon farm in Norway, it seems that no one said the salmon was really an elk—but bring in a Sami worker and that might change. Showing multiple salmon is less interesting than seeing how different salmon are made commensurable, or what sorts of salmon are only rarely enacted, and so on: the sorts of questions I like to ask come from the underlying program in social construction of technology. It would not be enough for us to say "gee there are multiple cars" because a car can be a form of transport, a washing machine, a chain saw, a tractor, and so on. What Ron Kline and I[66] tried to do with the work on the car is to show why some cars vanish and how and under what circumstances. The debate over ontology versus epistemology is a sideshow: the real issue—and here I find myself agreeing with much of what Woolgar and Lezaun,[67] Lynch,[68] and Sismondo[69] say on the matter—is how we can push on with science studies' long-term program of asking the "How could it be otherwise?" question and the politics of technology that follow from this.

ST Regarding long-term programs, I have a very last question. During the interview we've had the occasion to discuss the short and medium-term projects you are working on. Have you got any long-term plans?

TP Actually I do. You said to me in a conversation, "Trevor, with all European respect, I think you should put together a book where all your work is together." Actually, I've been talking with a publisher for a couple of years about doing such a book, which will be several of my case studies laced together with a SCOT material performativity-type framework overlaying them. So that is my long-term project: that's a project for a lifetime.

ST I am really looking forward to reading it! Thank you very much, Trevor.

66. Kline and Pinch 1996.
67. Woolgar and Lezaun 2013.
68. Lynch 2013.
69. Sismondo 2015.

References

Akrich, Madeleine. 1992a. Sémiotique et Sociologie Des Techniques: Jusqu'où Pousser Le Parallèle? In *Ces réseaux que la raison ignore: Centre de Sociologie de l'Innovation*, 24–30. Paris: L'Harmattan.

Akrich, Madeleine. 1992b. The de-scription of technical objects. In *Shaping Technology/Building Society: Studies in Sociotechnical Change*, ed. Wiebe E. Bijker and John Law, 205–225. Cambridge, MA: MIT Press.

Akrich, Madeleine, and Bruno Latour. 1992. A summary of a convenient vocabulary for the semiotics of human and nonhuman assemblies. In *Shaping Technology/Building Society: Studies in Sociotechnical Change*, ed. Wiebe E. Bijker and John Law, 259–264. Cambridge, MA: MIT Press.

Amin, Ash. 2000. Industrial districts. In *A Companion to Economic Geography*, ed. Eric Sheppard and Trevor J. Barnes, 149–168. Malden, MA: Blackwell.

Anderson, Ben, and Paul Harrison, eds. 2010. *Taking-Place: Non-Representational Theories and Geography*. Burlington, VT: Ashgate.

Appadurai, Arjun, ed. 1986. *The Social Life of Things: Commodities in Cultural Perspective*. Cambridge: Cambridge University Press.

Ashmore, Malcolm. 1989. *The Reflexive Thesis: Wrighting Sociology of Scientific Knowledge*. Chicago: University of Chicago Press.

Ashmore, Malcolm, Michael Joseph Mulkay, and Trevor J. Pinch. 1989. *Health and Efficiency: A Sociology of Health Economics*. Maidenhead, UK: Open University Press.

Aspers, Patrik. 2015. Performing ontology. *Social Studies of Science* 45 (3): 449–453.

Atkinson, J. Maxwell. 1984. *Our Masters' Voices: The Language and Body Language of Politics*. London: Methuen.

Atkinson, J. Maxwell, and Paul Drew. 1979. *Order in Court: The Organisation of Verbal Interaction in Judicial Settings*. London: Macmillan.

Bakardjieva, Maria. 2005. *Internet Society: The Internet in Everyday Life*. London: Sage.

Bal, Roland, Wiebe E. Bijker, and Ruud Hendriks. 2004. Democratisation of scientific advice. *British Medical Journal* 329 (7478): 1339–1341.

Barad, Karen M. 1998. Getting real: Technoscientific practices and the materialization of reality. *Differences* 10 (2): 87–126.

Barad, Karen M. 2007. *Meeting the Universe Halfway: Quantum Physics and the Entanglement of Matter and Meaning*. Durham: Duke University Press.

Barad, Karen M. 2011. Erasers and erasures: Pinch's unfortunate "uncertainty principle." *Social Studies of Science* 41 (3): 443–54.

Barley, Stephen R. 2015. Why the Internet makes buying a car less loathsome: How technologies change role relations. *Academy of Management Discoveries* 1 (1): 5–34.

Barnes, Barry. 1974. *Scientific Knowledge and Sociological Theory*. London: Routledge.

Barnes, Barry. 1977. *Interests and the Growth of Knowledge*. London: Routledge & Kegan Paul.

Barnes, Barry. 1982. *T. S. Kuhn and Social Science*. London: Macmillan.

Barnes, Barry, and R. G. A. Dolby. 1970. The scientific ethos: A deviant viewpoint. *Archives Européennes de Sociologie* 11:3–25.

Barnes, Barry, and Steven Shapin, eds. 1979. *Natural Order: Historical Studies of Scientific Culture*. London: Sage Publications.

Bell, Colin, and Helen Roberts, eds. 1984. *Social Researching: Politics, Problems, Practice*. London: Routledge & Kegan Paul.

Berthoin Antal, Ariane, Michael Hutter, and David Stark, eds. 2015. *Moments of Valuation: Exploring Sites of Dissonance*. Oxford: Oxford University Press.

Bijker, Wiebe E. 1987. The social construction of Bakelite: Toward a theory of invention. In *The Social Construction of Technological Systems: New Directions in the Sociology and History of Technology*, ed. Wiebe E. Bijker, Thomas Parke Hughes, and Trevor J. Pinch, 159–187. Cambridge, MA: MIT Press.

Bijker, Wiebe E. 1993. Do not despair: There is life after constructivism. *Science, Technology, and Human Values* 18:113–138.

Bijker, Wiebe E. 1995. *Of Bicycles, Bakelites, and Bulbs: Toward a Theory of Sociotechnical Change*. Cambridge, MA: MIT Press.

Bijker, Wiebe E. 2010. How is technology made? That is the question! *Cambridge Journal of Economics* 34 (1): 63–76.

Bijker, Wiebe E. 2015. Technology, social construction of. In *International Encyclopedia of the Social and Behavioral Sciences*, ed. James D. Wright, 135–140. Oxford: Elsevier.

Bijker, Wiebe E., Thomas Parke Hughes, and Trevor J. Pinch, eds. 1987. *The Social Construction of Technological Systems: New Directions in the Sociology and History of Technology*. Cambridge, MA: MIT Press.

Bijker, Wiebe E., and John Law, eds. 1992. *Shaping Technology/Building Society: Studies in Sociotechnical Change*. Cambridge, MA: MIT Press.

Bijker, Wiebe E., and Trevor J. Pinch. 2002. SCOT answers, other questions: A reply to Nick Clayton. *Technology and Culture* 43 (2): 361–369.

Bijsterveld, Karin. 2008. *Mechanical Sound: Technology, Culture, and Public Problems of Noise in the Twentieth Century*. Cambridge, MA: MIT Press.

Bijsterveld, Karin. Forthcoming. *Sonic Skills: Sound and Listening in Science, Engineering, and Medicine* .

Bijsterveld, Karin, Eefje Cleophas, Stefan Krebs, and Gijs Mom. 2014. *Sound and Safe: A History of Listening Behind the Wheel*. Oxford: Oxford University Press.

Bijsterveld, Karin, and Marten Schulp. 2004. Breaking into a world of perfection: Innovation in today's classical musical instruments. *Social Studies of Science* 34 (5): 649–674.

Bloor, David. 1973. Wittgenstein and Mannheim on the sociology of mathematics. *Studies in History and Philosophy of Science* 4 (2): 173–191.

Bloor, David. 1975. A philosophical approach to science. *Social Studies of Science* 5 (4): 507–517.

Bloor, David. 1976. *Knowledge and Social Imagery*. London: Routledge & Kegan Paul.

Bloor, David. 2003. Obituary: David Owen Edge (4 September 1932–28 January 2003). *Social Studies of Science* 33 (2): 171–176.

Bloor, David. 2008. Relativism at 30,000 feet. In *Knowledge as Social Order: Rethinking the Sociology of Barry Barnes*, ed. Massimo Mazzotti, 13–24. Burlington, VT: Ashgate.

Boczkowski, Pablo, and Leah A. Lievrouw. 2007. Bridging STS and communication studies: Scholarship on media and information technologies. In *The Handbook of Science and Technology Studies*, ed. by Edward J Hackett, Olga Amsterdamska, Michael E. Lynch, and Judy Wajcman, 949–978. Cambridge, MA: MIT Press.

Bowker, Geoffrey C., and Susan Leigh Star. 1999. *Sorting Things Out: Classification and Its Consequences*. Cambridge, MA: MIT Press.

Bricmont, Jean, and Alan D. Sokal. 2001. Remarks on methodological relativism and "antiscience." In *The One Culture? A Conversation about Science*, ed. Jay A. Labinger and H. M. Collins, 179–183. Chicago: University of Chicago Press.

Bull, Michael. 2000. *Sounding Out the City: Personal Stereos and the Management of Everyday Life*. Oxford: Berg.

Bull, Michael. 2008. *Sound Moves: iPod Culture and Urban Experience*. London: Routledge.

Bull, Michael, and Les Back, eds. 2015. *The Auditory Culture Reader*, 2nd ed. London: Bloomsbury Academic.

Callon, Michel. 1986. Éléments pour une sociologie de la Traduction: La domestication des coquilles Saint-Jacques et des marins-pêcheurs dans la baie de Saint-Brieuc. *L'Annee Sociologique* 36:169–208.

Callon, Michel. 1987. Society in the making: The study of technology as a tool for sociological analysis. In *The Social Construction of Technological Systems: New Directions in the Sociology and History of Technology*, ed. Wiebe E. Bijker, Thomas Parke Hughes, and Trevor J. Pinch, 83–103. Cambridge, MA: MIT Press.

Callon, Michel. 1999. Actor-network theory: The market test. In *Actor Network Theory and After*, ed. John Law and John Hassard, 181–195. Malden, MA: Blackwell.

Callon, Michel, and Bruno Latour. 1992. Don't throw the baby out with the Bath school! A reply to Collins and Yearley. In *Science as Practice and Culture*, ed. Andrew Pickering, 343–368. Chicago: University of Chicago Press.

Callon, Michel, John Law, and Arie Rip. 1986. Putting texts in their places. In *Mapping the Dynamics of Science and Technology: Sociology of Science in the Real World*, ed. Michel Callon, John Law, and Arie Rip, 221–230. London: Macmillan.

Castells, Manuel. 2001. *The Internet Galaxy: Reflections on the Internet, Business, and Society*. Oxford: Oxford University Press.

Cicourel, Aaron Victor. 1964. *Method and Measurement in Sociology*. New York: Free Press.

Cicourel, Aaron Victor. 1974. *Cognitive Sociology: Language and Meaning in Social Interaction*. New York: Free Press.

Clark, Colin, Paul Drew, and Trevor J. Pinch. 1994. Managing customer "objections" during real-life sales negotiations. *Discourse and Society* 5 (4): 437–462.

Clark, Colin, Peter Drew, and Trevor J. Pinch. 2003. Managing prospect affiliation and rapport in real-life sales encounters. *Discourse Studies* 5 (1): 5–31.

Clark, Colin, and Trevor J. Pinch. 1992. The anatomy of a deception: Fraud and finesse in the mock auction sales "con." *Qualitative Sociology* 15 (2): 151–175.

Clark, Colin, and Trevor J. Pinch. 1994. The interactional study of exchange rela-
tionships: An analysis of patter merchants at work on street markets. In *Higgling
Transactors and Their Markets in the History of Economics*, ed. E. Neil De Marchi and
Mary S. Morgan. Durham: Duke University Press.

Clark, Colin, and Trevor J. Pinch. 1995. *The Hard Sell: The Language and Lessons of
Street-Wise Marketing*. London: HarperCollins.

Clark, Colin, and Trevor J. Pinch. 2014. *The Hard Sell: The Language and Lessons of
Streetwise Marketing*. London: Sociografica.

Clayman, Steven, and John Heritage. 2002. *The News Interview: Journalists and Public
Figures on the Air*. Cambridge: Cambridge University Press.

Cockburn, Cynthia, and Susan Ormrod. 1993. *Gender and Technology in the Making*.
London: Sage.

Collin, Finn. 2010. *Science Studies as Naturalized Philosophy*, vol. 348. Berlin: Springer.

Collins, Harry M. 1974. The TEA set: Tacit knowledge and scientific networks. *Social
Studies of Science* 4 (2): 165–185.

Collins, Harry M. 1975. The seven sexes: A study in the sociology of a phenomenon,
or the replication of experiments in physics. *Sociology* 9:205–224.

Collins, Harry M. 1979. The investigation of frames of meaning in science: Comple-
mentarity and compromise. *Sociological Review* 27 (4): 703–718.

Collins, Harry M. 1981a. Knowledge and controversy: Studies of modern natural
science. [Special issue] *Social Studies of Science* 11:1–158.

Collins, Harry M. 1981b. Son of seven sexes: The social destruction of a physical
phenomenon. *Social Studies of Science* 11 (1): 33–62.

Collins, Harry M. 1981c. Stages in the Empirical Programme of Relativism. *Social
Studies of Science* 11 (1): 3–10.

Collins, Harry M. 1981d. What is TRASP? The radical programme as a methodological imperative. *Philosophy of the Social Sciences* 12 (2): 215.

Collins, Harry M. 1983a. An empirical relativist programme in the sociology of
scientific knowledge. In *Science Observed: Perspectives on the Social Study of Science*,
ed. Karin Knorr-Cetina and Michael Joseph Mulkay, 85–114. London: Sage.

Collins, Harry M. 1983b. The meaning of lies: Accounts of action and participatory
research. In *Accounts and Action: Surrey Conferences on Sociological Theory and Method*,
ed. G. Nigel Gilbert and Peter Abell, 69–78. Aldershot: Gower.

Collins, Harry M. 1983c. The sociology of scientific knowledge: Studies of contemporary science. *Annual Review of Sociology* 9:265–285.

Collins, Harry M. 1984. Researching spoonbending: Concepts and methods of participatory fieldwork. In *Social Researching: Politics, Problems, Practice*, ed. Colin Bell and Helen Roberts, 54–69. London: Routledge & Kegan Paul.

Collins, Harry M. 1985. *Changing Order: Replication and Induction in Scientific Practice*. London: Sage.

Collins, Harry M. 1990. *Artificial Experts: Social Knowledge and Intelligent Machines*. Inside Technology series. Cambridge, MA: MIT Press.

Collins, Harry M. 2002. The experimenter's regress as philosophical sociology. *Studies in History and Philosophy of Science, Part A* 33 (1): 149–156.

Collins, Harry M. 2003. Lead into gold: The science of finding nothing. *Studies in History and Philosophy of Science, Part A* 34 (4): 661–691.

Collins, Harry M. 2004. *Gravity's Shadow: The Search for Gravitational Waves*. Chicago: University of Chicago Press.

Collins, Harry M. 2011. *Tacit and Explicit Knowledge*. Chicago: University of Chicago Press.

Collins, Harry M. 2014. *Gravity's Ghost and Big Dog: Scientific Discovery and Social Analysis in the Twenty-First Century*. Chicago: University of Chicago Press.

Collins, Harry M., and Robert Evans. 2002. The third wave of science studies: Studies of expertise and experience. *Social Studies of Science* 32 (2): 235–296.

Collins, Harry M., and Robert Evans. 2007. *Rethinking Expertise*. Chicago: University of Chicago Press.

Collins, Harry M., and R. G. Harrison. 1975. Building a TEA laser: The caprices of communication. *Social Studies of Science* 5 (4): 441–450.

Collins, Harry M., and Trevor J. Pinch. 1979. The construction of the paranormal: Nothing unscientific is happening. In *On the Margins of Science: The Social Construction of Rejected Knowledge*, ed. Roy Wallis, 237–270. Keele: University of Keele.

Collins, Harry M., and Trevor J. Pinch. 1982. *Frames of Meaning: The Social Construction of Extraordinary Science*. London: Routledge & Kegan Paul.

Collins, Harry M., and Trevor J. Pinch. 1993. *The Golem: What You Should Know about Science*. Cambridge: Cambridge University Press.

Collins, Harry M., and Trevor J. Pinch. 1998. *The Golem at Large: What You Should Know about Technology*. Cambridge: Cambridge University Press.

Collins, Harry M., and Trevor J. Pinch. 2005. *Dr. Golem: How to Think about Medicine*. Chicago: University of Chicago Press.

Collins, Harry M., and Trevor J. Pinch. 2007. Who is to blame for the *Challenger* explosion? *Studies in History and Philosophy of Science, Part A* 38 (1): 254–255.

Collins, Harry M., and Martin Weinel. 2011. Transmuted expertise: How technical non-experts can assess experts and expertise. *Argumentation* 25 (3): 401–413.

Collins, Harry M., Martin Weinel, and Robert Evans. 2010. The politics and policy of the third wave: New technologies and society. *Critical Policy Studies* 4 (2): 185–201.

Collins, Harry M., and Steven Yearley. 1992. Epistemological chicken. In *Science as Practice and Culture*, ed. Andrew Pickering, 301–326. Chicago: University of Chicago Press.

Conant, James Bryant, ed. 1948. *Harvard Case Histories in Experimental Science*, vols. 1–2. Cambridge, MA: Harvard University Press.

Couldry, Nick. 2008. Actor network theory and media: Do they connect and on what terms? In *Connectivity, Networks, and Flows: Conceptualizing Contemporary Communications*, ed. Andreas Hepp, Friedrich Krotz, Shaun Moores, and Carsten Winter, 93–110. Cresskill: Hampton Press.

Cowan, Ruth Schwartz. 1983. *More Work for Mother: The Ironies of Household Technology from the Open Hearth to the Microwave*. New York: Basic Books.

Crane, Diana. 1965. Scientists at major and minor universities: A study of productivity and recognition. *American Sociological Review* 30:699–714.

Crane, Diana. 1969. Social structure in a group of scientists: A test of the "invisible college" hypothesis. *American Sociological Review* 36:335–352.

Crane, Diana. 1972. *Invisible Colleges: Diffusion of Knowledge in Scientific Communities*. Chicago: University of Chicago Press.

Culler, Jonathan D. 1982. *On Deconstruction: Theory and Criticism after Structuralism*. Ithaca: Cornell University Press.

Darr, Asaf. 2006. *Selling Technology: The Changing Shape of Sales in an Information Economy*. Ithaca: Cornell University Press.

Darr, Asaf, and Trevor J. Pinch. 2013. Performing sales: Material scripts and the social organization of obligation. *Organization Studies* 34 (11): 1601–1621.

David, Shay, and Trevor J. Pinch. 2006. Six degrees of reputation: The use and abuse of online review and recommendation systems. *First Monday* 11 (6). http://firstmonday.org/ojs/index.php/fm/article/view/1590/1505.

Dolby, R. G. A. 1974. In defence of a social criterion of scientific objectivity. *Science Studies* 4:187–190.

Drew, Paul, and John Heritage. 1992. *Talk at Work: Interaction in Institutional Settings*. Cambridge: Cambridge University Press.

Drew, Paul, and John Heritage. 2013. *Contemporary Studies in Conversation Analysis*. London: Sage Publications.

Duhem, Pierre Maurice Marie. 1954. *The Aim and Structure of Physical Theory*. [La Théorie physique: son objet, sa structure, 1906.] Princeton: Princeton University Press.

Eco, Umberto. 1992. *Interpretation and Overinterpretation*. Ed. Stefan Collini. Cambridge: Cambridge University Press.

Eco, Umberto. 2000. *Kant and the Platypus: Essays on Language and Cognition*. Boston: Houghton Mifflin Harcourt.

Edge, David. 1971. Science studies: Portrait of a course. *Technology and Society* 6:84–87.

Edge, David. 1976. Quantitative measures of communication in science. Paper presented at the International Symposium on Quantitative Measures in the History of Science, Berkeley, California, Aug. 25–27, 1976.

Edwards, Derek, Malcolm Ashmore, and Jonathan Potter. 1995. Death and furniture: The rhetoric, politics, and theology of bottom line arguments against relativism. *History of the Human Sciences* 8 (2): 25–49.

Edwards, Paul N. 1996. *The Closed World: Computers and the Politics of Discourse in Cold War America*. Cambridge, MA: MIT Press.

Edwards, Paul N. 2010. *A Vast Machine: Computer Models, Climate Data, and the Politics of Global Warming*. Cambridge, MA: MIT Press.

Elliott, Brian, ed. 1988. *Technology and Social Change*. Edinburgh: Edinburgh University Press.

Epstein, Steven. 1996. *Impure Science: AIDS, Activism, and the Politics of Knowledge*. Berkeley: University of California Press.

Feenberg, Andrew. 2006. Symmetry, asymmetry, and the real possibility of radical change: Reply to Kochan. *Studies in History and Philosophy of Science* 37 (4): 721–727.

Festinger, Leon, Henry W. Riecken, and Stanley Schachter. 1956. *When Prophecy Fails*. Minneapolis: University of Minnesota Press.

Feyerabend, Paul. 1993. *Against Method*, 3rd ed. London: Verso. (1st ed. 1975.)

Fish, Stanley Eugene. 1980. *Is There a Text in This Class? The Authority of Interpretive Communities*. Cambridge, MA: Harvard University Press.

Fish, Stanley Eugene. 1996. Professor Sokal's bad joke. *New York Times*, May 21.

Garfinkel, Harold. 1967. *Studies in Ethnomethodology*. Englewood Cliffs, NJ: Prentice-Hall.

Garfinkel, Harold, Michael Lynch, and Eric Livingston. 1981. The work of a discovering science construed with materials from the optically discovered pulsar. *Philosophy of the Social Sciences* 11:131–158.

Georgescu-Roegen, Nicholas. 1971. *The Entropy Law and the Economic Process.* Cambridge, MA: Harvard University Press.

Gilbert, G. Nigel. 1976a. The development of science and scientific knowledge: The case of radar meteor research. In *Perspectives on the Emergence of Scientific Disciplines,* ed. Gérard Lemaine, M. MacLeod, Michael J. Mulkay, and Peter Weingart, 187–206. Paris: Mouton.

Gilbert, G. Nigel. 1976b. The transformation of research findings into scientific knowledge. *Social Studies of Science* 6:281–306.

Gilbert, G. Nigel. 1977. Referencing as persuasion. *Social Studies of Science* 7 (1): 113–122.

Gilbert, G. Nigel, and Peter Abell, eds. 1983. *Accounts and Action: Surrey Conferences on Sociological Theory and Method.* Aldershot: Gower.

Gilbert, G. Nigel, and Michael J. Mulkay. 1980. Contexts of scientific discourse: Social accounting in experimental papers. In *The Social Process of Scientific Investigation,* ed. Karin Dagmar Knorr, R. Krohn, and Richard Whitley, 269–296. Reidel: Dordrecht.

Gilbert, G. Nigel, and Michael J. Mulkay. 1984. *Opening Pandora's Box: A Sociological Analysis of Scientists' Discourse.* Cambridge: Cambridge University Press.

Gilbert, G. Nigel, and Steve Woolgar. 1974. The quantitative study of science: An examination of the literature. *Science Studies* 4 (3): 279–294.

Gitelman, Lisa. 2006. *Always Already New: Media, History and the Data of Culture.* Cambridge, MA: MIT Press.

Goffman, Erving. 1959. *The Presentation of Self in Everyday Life.* Garden City, NY: Doubleday.

Gooding, David, Trevor J. Pinch, and Simon Schaffer, eds. 1989. *The Uses of Experiment: Studies in the Natural Sciences.* Cambridge: Cambridge University Press.

Goodman, Steve. 2010. *Sonic Warfare: Sound, Affect, and the Ecology of Fear.* Cambridge, MA: MIT Press.

Greenberg, Joshua M. 2010. *From Betamax to Blockbuster: Video Stores and the Invention of Movies on Video.* Cambridge, MA: MIT Press.

Groeneveld, L., N. Koller, and Nicholas C. Mullins. 1975. The Advisers of the U.S. National Science Foundation. *Social Studies of Science* 5 (3): 343–354.

Gross, Paul R., and N. Levitt. 1994. *Higher Superstition: The Academic Left and Its Quarrels with Science*. Baltimore: The Johns Hopkins University Press.

Hackett, Edward J., Olga Amsterdamska, Michael E. Lynch, and Judy Wajcman, eds. 2007. *The Handbook of Science and Technology Studies*. Cambridge, MA: MIT Press.

Hanson, Norwood Russell. 1958. *Patterns of Discovery: An Inquiry into the Conceptual Foundations of Science*. Cambridge: Cambridge University Press.

Hanson, Norwood Russell. 1969. *Perception and Discovery: An Introduction to Scientific Inquiry*. San Francisco: Freeman, Cooper.

Harding, Sandra G. 1986. *The Science Question in Feminism*. Ithaca: Cornell University Press.

Harding, Sandra G. 1998. *Is Science Multicultural? Postcolonialisms, Feminisms, and Epistemologies (Race, Gender, and Science)*. Bloomington: Indiana University Press.

Harding, Sandra G., ed. 2011. *The Postcolonial Science and Technology Studies Reader*. Durham: Duke University Press.

Harvey, Bill. 1981. Plausibility and the evaluation of knowledge: A case-study of experimental quantum mechanics. *Social Studies of Science* 11 (1): 95–130.

Hecht, Gabrielle. 1998. *The Radiance of France: Nuclear Power and National Identity after World War II*. Cambridge, MA: MIT Press.

Hennion, Antoine. 2015. Paying attention: What is tasting wine about? In *Moments of Valuation: Exploring Sites of Dissonance*, ed. Ariane Berthoin Antal, Michael Hutter and David Stark, 37–56. Oxford: Oxford University Press.

Henry, J. 2008. Historical and other studies of science, technology, and medicine in the University of Edinburgh. *Notes and Records of the Royal Society* 62 (2): 223–235.

Hess, David J. 1993. *Science in the New Age: The Paranormal, Its Defenders and Debunkers, and American Culture*. Madison: University of Wisconsin Press.

Hess, David J. 1996. Technology and alternative cancer therapies: An analysis of heterodoxy and constructivism. *Medical Anthropology Quarterly: International Journal for the Cultural and Social Analysis of Health* 10:657–674.

Hess, David J. 1997. *Science Studies: An Advanced Introduction*. New York: NYU Press.

Hobsbawm, Eric J., and Terence O. Ranger. 1983. *The Invention of Tradition*. Cambridge: Cambridge University Press.

Holton, Gerald. 1973. *Thematic Origins of Scientific Thought: Kepler to Einstein*. Cambridge, MA: Harvard University Press.

Holton, Gerald. 1978. *The Scientific Imagination: Case Studies*. Cambridge: Cambridge University Press.

Holton, Gerald. 1986. *The Advancement of Science, and Its Burdens: The Jefferson Lecture and Other Essays*. Cambridge: Cambridge University Press.

Horning, Susan Schmidt. 2004. Engineering the performance: Recording engineers, tacit knowledge, and the art of controlling sound. *Social Studies of Science* 34 (5): 703–731.

Hughes, Thomas Parke. 1986. The seamless web: Technology, science, etcetera, etcetera. *Social Studies of Science, 16* (2): 281–292.

Hughes, Thomas Parke. 1993. *Networks of Power: Electrification in Western Society, 1880–1930*. Baltimore: The Johns Hopkins University Press.

Hughes, Thomas Parke. 2012. The evolution of large technological systems. In *The Social Construction of Technological Systems: New Directions in the Sociology and History of Technology*, ed. Wiebe E. Bijker, Thomas Parke Hughes, and Trevor J. Pinch, 51–82. Cambridge, MA: MIT Press. (1st. ed. 1987.)

Humphreys, Lee. 2005. Reframing social groups, closure, and stabilization in the social construction of technology. *Social Epistemology* 19 (2–3): 231–253.

Ihde, D. 1976. *Listening and Voice: A Phenomenology of Sound*. Athens, OH: Ohio University Press.

Jackson, Michèle H., Marshall Scott Poole, and Tim Kuhn. 2002. The social construction of technology in studies of the workplace. In *Handbook of New Media: Social Shaping and Consequences of ICTs*, ed. Leah A Lievrouw and Sonia M Livingstone, 236–253. London: Sage.

Jarzabkowski, Paula, and Trevor J. Pinch. 2013. Sociomateriality is "the new black": Accomplishing repurposing, reinscripting, and repairing in context. *Management* 16 (5): 579–592.

Jasanoff, Sheila, ed. 2004. *States of Knowledge: The Co-Production of Science and the Social Order*. London: Routledge.

Kessous, Emmanuel, and Alexandre Mallard, eds. 2014. *La Fabrique de La Vente: Le Travail Commercial Dans Les Télécommunications*. Paris: Presses des Mines.

Kitcher, Phil. 1994. How the road to relativism is paved. Paper presented at the HSS Annual Meeting, New Orleans, Oct. 12.

Kline, Ronald, and Trevor Pinch. 1996. Users as agents of technological change: The social construction of the automobile in the rural United States. *Technology and Culture* 37 (4): 763.

Knappett, Carl, and Lambros Malafouris, eds. 2008. *Material Agency: Towards a Non-Anthropocentric Approach*. Berlin: Springer.

Knorr, Karin. 1977. Producing and reproducing knowledge: Descriptive or constructive? Toward a model of research production. *Social Sciences Information: Information Sur les Sciences Sociales* 16:669–696.

Knorr, Karin. 1979. Tinkering toward success: Prelude to a theory of scientific practice. *Theory and Society* 8 (3): 347–376.

Knorr, Karin, Barbara R. Krohn, and Richard Whitley, eds. 1981. *The Social Process of Scientific Investigation*. Berlin: Springer.

Knorr-Cetina, Karin. 1981a. Social and scientific method, or What do we make of the distinction between the natural and the social sciences? *Philosophy of the Social Sciences* 11 (3): 335.

Knorr-Cetina, Karin. 1981b. *The Manufacture of Knowledge: An Essay on the Constructivist and Contextual Nature of Science*. Oxford: Pergamon Press.

Knorr-Cetina, Karin. 1983. The ethnographic study of scientific work: Towards a constructivist interpretation of science. In *Science Observed: Perspectives on the Social Study of Science*, ed. Karin Knorr-Cetina and Michael Joseph Mulkay, 115–140. London: Sage Publications.

Knorr-Cetina, Karin, and Michael Joseph Mulkay, eds. 1983. *Science Observed: Perspectives on the Social Study of Science*. London: Sage.

Kuhn, Thomas S. 1970. *The Structure of Scientific Revolutions*. Chicago: University of Chicago Press.

Kuhn, Thomas S. 1978. *Black-Body Theory and the Quantum Discontinuity, 1894–1912*. Chicago: University of Chicago Press.

LaBelle, Brandon. 2010. *Acoustic Territories: Sound Culture and Everyday Life*. New York: Continuum Books.

Labinger, Jay A., and H. M. Collins, eds. 2001. *The One Culture? A Conversation about Science*. Chicago: University of Chicago Press.

Lakatos, Imre. 1976. *Proofs and Refutations: The Logic of Mathematical Discovery*. Cambridge: Cambridge University Press.

Lakatos, Imre. 1978. *The Methodology of Scientific Research Programmes*, vol. 1: *Philosophical Papers*. Cambridge: Cambridge University Press.

Lakatos, Imre. 1980. *Mathematics, Science, and Epistemology*, vol. 2: *Philosophical Papers*. Cambridge: Cambridge University Press.

Latour, Bruno. 1986. Visualization and cognition. *Knowledge in Society* 6:1–40.

Latour, Bruno. 1988a. A relativistic account of Einstein's relativity. *Social Studies of Science* 18 (1): 3–44.

Latour, Bruno. 1988b. How to write "The Prince" for machines as well as for machinations. In *Technology and Social Change*, ed. Brian Elliott, 20–43. Edinburgh: Edinburgh University Press.

Latour, Bruno. 1992. Where are the missing masses? The sociology of a few mundane artifacts. In *Shaping Technology/Building Society: Studies in Sociotechnical Change*, ed. Wiebe E. Bijker and John Law, 225–258. Cambridge, MA: MIT Press.

Latour, Bruno. 1993. *We Have Never Been Modern*. Cambridge, MA: Harvard University Press.

Latour, Bruno. 1996. *Aramis, or the Love of Technology*. Cambridge, MA: Harvard University Press.

Latour, Bruno. 1999. *Pandora's Hope: Essays on the Reality of Science Studies*. Cambridge, MA: Harvard University Press.

Latour, Bruno. 2005. *Reassembling the Social: An Introduction to Actor-Network-Theory*. Oxford: Oxford University Press.

Latour, Bruno, and Emilie Hermant. 1998. *Paris, ville invisible*. Paris: Le Plessis-Robinson: Empecheurs.

Latour, Bruno, and Couze Venn. 2002. Morality and technology: The end of the means. *Theory, Culture, and Society* 19 (5–6): 247–60.

Latour, Bruno, and Steve Woolgar. 1979. *Laboratory Life: The Construction of Scientific Facts*. Princeton: Princeton University Press.

Lave, Jean, and Etienne Wenger. 1991. *Situated Learning: Legitimate Peripheral Participation*. Cambridge: Cambridge University Press.

Lave, Jean, and Etienne Wenger. 2010. Care and killing: Tensions in veterinary practice. In *Care in Practice: On Tinkering in Clinics, Homes, and Farms*, ed. Annemarie Mol, Ingunn Moser and Jeanette Pols, 57–69. Bielefeld: Transcript.

Law, John. 1987. Technology and heterogeneous engineering: The case of Portuguese expansion. In *The Social Construction of Technological Systems: New Directions in the Sociology and History of Technology*, ed. Wiebe E. Bijker, Thomas Parke Hughes, and Trevor J. Pinch, 111–134. Cambridge, MA: MIT Press.

Law, John, and Michel Callon. 1992. The life and death of an aircraft: A network analysis of technical change. In *Shaping Technology/Building Society: Studies in Sociotechnical Change*, ed. Wiebe E. Bijker and John Law, 21–52. Cambridge, MA: MIT Press.

Law, John, and John Hassard, eds. 1999. *Actor Network Theory and After*. Oxford: Blackwell.

Law, John, and Marianne Elisabeth Lien. 2013. Slippery: Field notes in empirical ontology. *Social Studies of Science* 43 (3): 363–378.

Law, John, and Annemarie Mol. 2008. The actor-enacted: Cumbrian sheep in 2001. In *Material Agency: Towards a Non-Anthropocentric Approach*, ed. Carl Knappett and Lambros Malafouris, 57–77. Berlin: Springer.

Lemaine, Gérard, M. MacLeod, Michael J. Mulkay, and Peter Weingart, eds. 1976. *Perspectives on the Emergence of Scientific Disciplines*. Paris: Mouton.

Lievrouw, Leah A. 2002. Determination and contingency in new media development: Diffusion of innovation and social shaping of technology perspectives. In *Handbook of New Media: Social Shaping and Consequences of ICTs*, ed. Leah A. Lievrouw and Sonia M. Livingstone, 183–199. London: Sage.

Lievrouw, Leah A. 2014. Materiality and media in communication and technology studies: An unfinished project. In *Media Technologies: Essays on Communication, Materiality, and Society*, ed. Tarleton Gillespie, Pablo J. Boczkowski, and Kirsten A. Foot, 21–51. Cambridge, MA: MIT Press.

Lin, Ling-Fei. 2015. The dynamics of design-manufacturing laptops: How Taiwanese contract manufacturers matter in the history of laptop production. PhD dissertation, Cornell University, Ithaca, New York. http://ecommons.cornell.edu/handle/1813/39461 (last accessed July 7, 2015).

Luhmann, Niklas. 1979. *Trust and Power: Two Works*. Chichester: Wiley.

Lynch, Michael. 1982. Technical work and critical inquiry: Investigations in a scientific laboratory. *Social Studies of Science* 12 (4): 499–533.

Lynch, Michael. 1984. "Turning up signs" in neurobehavioral diagnosis. *Symbolic Interaction* 7 (1): 67–86.

Lynch, Michael. 1985a. *Art and Artifact in Laboratory Science: A Study of Shop Work and Shop Talk in a Research Laboratory*. London: Routledge & Kegan Paul.

Lynch, Michael. 1985b. Discipline and the material form of images: An analysis of scientific visibility. *Social Studies of Science* 15 (1): 37–66.

Lynch, Michael. 1993. *Scientific Practice and Ordinary Action: Ethnomethodology and Social Studies of Science*. Cambridge: Cambridge University Press.

Lynch, Michael. 2000. Against reflexivity as an academic virtue and source of privileged knowledge. *Theory, Culture, and Society* 17 (3): 26–54.

Lynch, Michael. 2013. Ontography: Investigating the production of things, deflating ontology. *Social Studies of Science* 43 (3): 444–462.

Lynch, Michael, Eric Livingston, and Harold Garfinkel. 1983. Temporal order in laboratory work. In *Science Observed: Perspectives on the Social Study of Science*, ed. Karin Knorr-Cetina and Michael Joseph Mulkay. London: Sage.

Lyon, David. 2002. *Surveillance Society: Monitoring Everyday Life.* Buckingham: Open University Press.

Lyon, David. 2003. *Surveillance after September 11.* Malden, MA: Polity Press, in association with Blackwell.

Mackay, Hughie, and Gareth Gillespie. 1992. Extending the social shaping of technology approach: Ideology and appropriation. *Social Studies of Science* 22 (4): 685–716.

MacKenzie, Donald A. 1981a. Sociobiologies in competition: The biometrician–Mendelian controversy. In *Biology, Medicine, and Society, 1840–1940,* ed. Charles Webster, 243–288. Cambridge: Cambridge University Press.

MacKenzie, Donald A. 1981b. *Statistics in Britain, 1865–1930: The Social Construction of Scientific Knowledge.* Edinburgh: Edinburgh University Press.

MacKenzie, Donald A. 1989. From Kwajalein to Armageddon? Testing and the social construction of missile accuracy. In *The Uses of Experiment: Studies in the Natural Sciences,* ed. David Gooding, Trevor J. Pinch, and Simon Schaffer, 409–436. Cambridge: Cambridge University Press.

MacKenzie, Donald A. 2008. What's in a number? *London Review of Books* 30 (18): 11–12. http://www.lrb.co.uk/v30/n18/donald-mackenzie/whats-in-a-number.

MacKenzie, Donald A., and Barry Barnes. 1979. Scientific judgment: The biometry–Mendelism controversy. In *Natural Order: Historical Studies of Scientific Culture,* ed. Barry Barnes and Steven Shapin, 191–210. London: Sage.

MacKenzie, Donald A., and Graham Spinardi. 1988a. The shaping of nuclear weapon system technology: U.S. fleet ballistic missile guidance and navigation: I: From Polaris to Poseidon. *Social Studies of Science* 18 (3): 419–463.

MacKenzie, Donald A., and Graham Spinardi. 1988b. The shaping of nuclear weapon system technology: US fleet ballistic missile guidance and navigation: II: "Going for broke"—The path to Trident II. *Social Studies of Science* 18 (4): 581–624.

MacKenzie, Donald A., and Judy Wajcman, eds. 1985. *The Social Shaping of Technology: How the Refrigerator Got Its Hum.* Milton Keynes: Open University Press.

Magaudda, Paolo. 2014. The broken boundaries between science and technology studies and cultural sociology: Introduction to an interview with Trevor Pinch. *Cultural Sociology* 8 (1): 63–76.

Manzarek, Ray. 1998. *Light My Fire: My Life with the Doors.* New York: Putnam.

Marshall, Alfred. 1919. *Industry and Trade.* London: Macmillan.

Mazzotti, Massimo, ed. 2008. *Knowledge as Social Order: Rethinking the Sociology of Barry Barnes.* Aldershot: Ashgate.

Mendelsohn, Everett, Peter Weingart, and Richard Whitley, eds. 1977. *The Social Production of Scientific Knowledge.* Dordrecht: D. Reidel.

Mermin, David. 1996. What's wrong with this sustaining myth? *Physics Today* 49 (3): 11.

Mermin, David. 1997. What's wrong with this reading? *Physics Today* 50 (10): 11–13.

Merton, Robert K. 1937. The sociology of knowledge. *Isis* 27 (3): 493–503.

Merton, Robert K. 1938. Science, technology, and society in seventeenth century England. *Osiris* 4:360–632.

Merton, Robert K. 1957. *Social Theory and Social Structure.* New York: Free Press.

Merton, Robert K. 1973. *The Sociology of Science: Theoretical and Empirical Investigations.* Chicago: University of Chicago Press.

Mills, Mara. 2012. Do signals have politics? Inscribing abilities in cochlear implants. In *The Oxford Handbook of Sound Studies,* ed. Trevor J. Pinch and Karin Bijsterveld, 320–346. New York: Oxford University Press.

Mitroff, I. 1974. *The Subjective Side of Science.* Amsterdam: Elsevier.

Mol, Annemarie. 2002. *The Body Multiple: Ontology in Medical Practice.* Durham, NC: Duke University Press.

Mol, Annemarie, Ingunn Moser, and J. Pols, eds. 2010. *Care in Practice: On Tinkering in Clinics, Homes, and Farms.* Bielefeld; Piscataway, NJ: Transcript, distributed in North America by Transaction Publishers.

Moores, Shaun. 2012. *Media, Place, and Mobility.* New York: Palgrave Macmillan.

Mulkay, Michael J. 1969. Some aspects of cultural growth in the natural sciences. *Social Research* 36:22–52.

Mulkay, Michael J. 1976. Norms and ideology in science. *Social Science Information* 15 (4–5): 637–656.

Mulkay, Michael J. 1979. *Science and the Sociology of Knowledge.* London: G. Allen & Unwin.

Mulkay, Michael J. 1981. Action and belief or scientific discourse? A possible way of ending intellectual vassalage in social studies of science. *Philosophy of the Social Sciences* 11:163–171.

Mulkay, Michael J. 1984. The scientist talks back: A one-act play, with a moral, about replication in science and reflexivity in sociology. *Social Studies of Science* 14 (2): 265–283.

Mulkay, Michael J., and G. Nigel Gilbert. 1982. What is the ultimate question? Some remarks in defence of the analysis of scientific discourse. *Social Studies of Science* 12 (2): 309–319.

Mulkay, Michael J., and G. Nigel Gilbert. 1983. Scientists' theory talk. *Canadian Journal of Sociology* 8:179–197.

Mulkay, Michael J., G. Nigel Gilbert, and Steve Woolgar. 1975. Problem areas and research networks in science. *Sociology* 9:187–203.

Mulkay, Michael J., Jonathan Potter, and Steven Yearley. 1983. Why an analysis of scientific discourse is needed. In *Science Observed: Perspectives on the Social Study of Science*, ed. Karin Knorr-Cetina and Michael J. Mulkay, 171–203. London: Sage.

Mullins, Nicholas C. 1972a. The development of a scientific specialty: The Phage Group and the origins of molecular biology. *Minerva* 10:51–82.

Mullins, Nicholas C. 1972b. The structure of an elite: The advisory structure of the U.S. Public Health Service. *Science Studies* 2:3–29.

Mullins, Nicholas C. 1973. The development of specialties in social science: The case of ethnomethodology. *Science Studies* 3:245–273.

Mullins, Nicholas C., L. Hargens, P. K. Hetch, and E. L. Kick. 1977. The group structure of co-citation clusters: A comparative study. *American Sociological Review* 42:552–562.

Myers, Greg. 1990. *Writing Biology: Texts in the Social Construction of Scientific Knowledge*. Madison: University of Wisconsin Press.

Nowotny, Helga, and Hilary Rose, eds. 1979. *Counter-Movements in the Sciences: The Sociology of the Alternatives to Big Science*. Dordrecht: D. Reidel.

Oudshoorn, Nelly. 2003. *The Male Pill: A Biography of a Technology in the Making*. Durham: Duke University Press.

Oudshoorn, Nelly, and Trevor J. Pinch, eds. 2003. *How Users Matter: The Co-Construction of Users and Technologies*. Cambridge, MA: MIT Press.

Oudshoorn, Nelly, and Trevor J. Pinch, eds. 2007. User-technology relationships: Some recent developments. In *The Handbook of Science and Technology Studies*, ed. Edward J Hackett, Olga Amsterdamska, Michael E. Lynch, and Judy Wajcman, 541–565. Cambridge, MA: MIT Press.

Pamplin, Brian R., and Harry M. Collins. 1975. Spoon bending: An experimental approach. *Nature* 257 (5521): 8.

Pickering, Andrew. 1981. Constraints on controversy: The case of the magnetic monopole. *Social Studies of Science* 11 (1): 63–93.

Pickering, Andrew. 1992a. From science as knowledge to science as practice. In *Science as Practice and Culture*, ed. Andrew Pickering, 1–26. Chicago: University of Chicago Press.

Pickering, Andrew, ed. 1992b. *Science as Practice and Culture*. Chicago: University of Chicago Press.

Pickering, Andrew. 1995. *The Mangle of Practice: Time, Agency, and Science*. New York: Harper Torchbooks.

Pickering, Andrew, and Keith Guzik, eds. 2008. *The Mangle in Practice: Science, Society, and Becoming*. Durham, NC: Duke University Press.

Pinch, Trevor J. 1977. What does a proof do if it does not prove? In *The Social Production of Scientific Knowledge*, ed. Everett Mendelsohn, Peter Weingart, and Richard Whitley, 171–215. Dordrecht: D. Reidel.

Pinch, Trevor J. 1979a. Normal explanations of the paranormal: The demarcation problem and fraud in parapsychology. *Social Studies of Science* 9 (3): 329–348.

Pinch, Trevor J. 1979b. Paradigm lost? A review symposium. *Isis* 70 (3): 437–440.

Pinch, Trevor J. 1979c. The hidden-variables controversy in quantum physics. *Physics Education* 14 (1): 48–52.

Pinch, Trevor J. 1981a. The sun-set: The presentation of certainty in scientific life. *Social Studies of Science* 11 (1): 131–158.

Pinch, Trevor J. 1981b. Theoreticians and the Production of Experimental Anomaly: The Case of Solar Neutrinos. In *The Social Process of Scientific Investigation*, ed. Karin Knorr-Cetina, R. Khron, and Richard Child Whitley, 77–106. Berlin: Springer.

Pinch, Trevor J. 1982. Book review: *Science and the Sociology of Knowledge* by M. Mulkay. *4S Newsletter* 7 (2): 20–22.

Pinch, Trevor J. 1983. Reflecting on Reflexivity. *EASST Newsletter* 2 (2): 5–7.

Pinch, Trevor J. 1984. Kuhn and parapsychology. *Journal of Parapsychology* 48:121–125.

Pinch, Trevor J. 1985a. Recent trends in the history of technology. *BSHS Newsletter*, Jan. 16.

Pinch, Trevor J. 1985b. Theory testing in science: The case of solar neutrinos. *Philosophy of the Social Sciences* 15:167–187.

Pinch, Trevor J. 1985c. Towards an analysis of scientific observation: The externality and evidential significance of observational reports in physics. *Social Studies of Science* 15 (1): 3–36.

Pinch, Trevor J. 1986. *Confronting Nature: The Sociology of Solar-Neutrino Detection.* Berlin: Springer.

Pinch, Trevor J. 1987a. Some suggestions from sociology of science to advance the psi debate. *Behavioral and Brain Sciences* 10 (4): 603–5.

Pinch, Trevor J. 1987b. Book review: Art and artifact in laboratory science: A study of shop work and shop talk in a research laboratory. *Sociology of Health and Illness* 9 (2): 219–220.

Pinch, Trevor J. 1993a. "Testing—one, two, three … testing!" Toward a sociology of testing. *Science, Technology, and Human Values* 18 (1): 25–41.

Pinch, Trevor J. 1993b. Turn, turn, and turn again: The Woolgar formula. *Science, Technology, and Human Values* 18:511–522.

Pinch, Trevor J. 1996. Social construction of technology: A review. In *Technological Change: Methods and Themes in the History of Technology*, ed. Robert Fox, 17–36. Newark: Harwood Academic.

Pinch, Trevor J. 1997. Kuhn—the conservative and radical interpretations: Are some Mertonians "Kuhnians" and some Kuhnians "Mertonians"? *Social Studies of Science* 27 (3): 465–82.

Pinch, Trevor J. 1999. Mangled up in blue. *Studies in History and Philosophy of Science* 30:139–148.

Pinch, Trevor J. 2003. Giving birth to new users: How the Minimoog was sold to rock & roll. In *How Users Matter: The Co-Construction of Users and Technologies*, ed. Nelly Oudshoorn and Trevor J. Pinch, 247–270. Cambridge, MA: MIT Press.

Pinch, Trevor J. 2007. The synthesizer. In *Evocative Objects: Things We Think With*, ed. Sherry Turkle, 162–169. Cambridge, MA: MIT Press.

Pinch, Trevor J. 2008a. Relativism: Is it worth the candle? In *Knowledge as Social Order: Rethinking the Sociology of Barry Barnes*, ed. Massimo Mazzotti, 35–48. Aldershot: Ashgate.

Pinch, Trevor J. 2008b. Teaching sociology to science and engineering students: Some experiences from an introductory science and technology studies course. *Research in Social Problems and Public Policy* 16:99–114.

Pinch, Trevor J. 2008c. Technology and institutions: Living in a material world. *Theory and Society* 37 (5): 461–483.

Pinch, Trevor J. 2009. The social construction of technology (SCOT): The old, the new, and the nonhuman. In *Material Culture and Technology in Everyday Life: Ethnographic Approaches*, ed. Phillip Vannini. New York: Peter Lang.

Pinch, Trevor J. 2010a. On making infrastructure visible: Putting the non-humans to rights. *Cambridge Journal of Economics* 34 (1): 77–89.

Pinch, Trevor J. 2010b. The invisible technologies of Goffman's sociology from the merry-go-round to the Internet. *Technology and Culture* 51 (2): 409–424.

Pinch, Trevor J. 2011. Review essay: Karen Barad, quantum mechanics, and the paradox of mutual exclusivity. *Social Studies of Science* 41 (3): 431–441.

Pinch, Trevor J. 2014a. Immanuel Velikovsky and the return of the fringe. *Metascience* 23 (3): 525–529.

Pinch, Trevor J. 2014b. Space is the place: The electronic sounds of inner and outer space. *Journal of Sonic Studies* 8. http://www.researchcatalogue.net/view/108499/108500.

Pinch, Trevor J. 2015a. Moments in the valuation of sound: The early history of synthesizers. In *Moments of Valuation: Exploring Sites of Dissonance*, ed. Ariane Berthoin Antal, Michael Hutter, and David Stark, 15–36. Oxford: Oxford University Press.

Pinch, Trevor J. 2015b, forthcoming. The sound of economic exchange. In *The Auditory Culture Reader*, 2nd ed., ed. Michael Bull and Les Back. London: Bloomsbury Academic.

Pinch, Trevor J., Malcolm Ashmore, and Michael Mulkay. Social technologies: To test or not to test, that is the question. Paper presented at the International Workshop on the Integration of Social and Historical Studies of Technology, University of Twente, Sept. 3–5, 1987.

Pinch, Trevor J., Malcolm Ashmore, and Michael Mulkay. 1992. Technology, testing, text: Clinical budgeting in the U.K. National Health Service. In *Shaping Technology/Building Society: Studies in Sociotechnical Change*, ed. Wiebe E. Bijker and John Law, 265–289. Cambridge, MA: MIT Press.

Pinch, Trevor J., and Wiebe E. Bijker. 1984. The social construction of facts and artifacts: Or how the sociology of science and the sociology of technology might benefit each other. *Social Studies of Science* 14 (3): 399–441.

Pinch, Trevor J., and Wiebe E. Bijker. 1986. Science, relativism, and the new sociology of technology: Reply to Russell. *Social Studies of Science* 16 (2): 347–360.

Pinch, Trevor J., and Karin Bijsterveld. 2003. "Should one applaud?" Breaches and boundaries in the reception of new technology in music. *Technology and Culture* 44 (3): 536–559.

Pinch, Trevor J., and Karin Bijsterveld. 2004. Sound studies: New technologies and music. *Social Studies of Science* 34 (5): 635–648.

Pinch, Trevor J., and Karin Bijsterveld, eds. 2012. *The Oxford Handbook of Sound Studies*. New York: Oxford University Press.

Pinch, Trevor J., and Colin Clark. 1986. The hard sell: "Patter merchanting" and the strategic (re)production and local management of economic reasoning in the sales routines of market pitchers. *Sociology* 20 (2): 169–191.

Pinch, Trevor J., and Harry M. Collins. 1979. Is anti-science not-science? The case of parapsychology. In *Counter-Movements in the Sciences: The Sociology of the Alternatives to Big Science*, ed. Helga Nowotny and Hilary Rose, 221–250. Dordrecht: D. Reidel.

Pinch, Trevor J., and Harry M. Collins. 1984. Private science and public knowledge: The Committee for the Scientific Investigation of the Claims of the Paranormal and its use of the literature. *Social Studies of Science* 14 (4): 521–546.

Pinch, Trevor J., Harry M. Collins, and Larry Carbone. 1996. Inside knowledge: Second order measures of skills. *Sociological Review* 44 (2): 163–186.

Pinch, Trevor J., and Filip Kesler. 2011. How Aunt Ammy gets her free lunch: A study of the top-thousand customer reviewers at Amazon.com http://www .freelunch.me/ (last accessed Apr. 12, 2016).

Pinch, Trevor J., and Trevor J. Pinch. 1988. Reservations about reflexivity and new literary forms: Or why let the devil have all the good tunes? In *Knowledge and Reflexivity: New Frontiers in the Sociology of Knowledge*, ed. Steve Woolgar, 178–197. London: Sage.

Pinch, Trevor J., and Frank Trocco. 1998. The social construction of the electronic music synthesizer. *ICON: Journal of the International Committee for the History of Technology* 4:67–83.

Pinch, Trevor J., and Frank Trocco. 2002. *Analog Days: The Invention and Impact of the Moog Synthesizer*. Cambridge, MA: Harvard University Press.

Polanyi, Michael. 1958. *Personal Knowledge: Towards a Post-Critical Philosophy*. London: Routledge.

Polanyi, Michael. 1966. *The Tacit Dimension*. Garden City, NY: Doubleday.

Popper, Karl Raimund. 1959. *The Logic of Scientific Discovery*. New York: Harper & Row.

Popper, Karl Raimund. 1962. *Conjectures and Refutations: The Growth of Scientific Knowledge*. London: Routledge.

Popper, Karl Raimund. 1972. *Objective Knowledge: An Evolutionary Approach*. Oxford: Clarendon Press.

Porcello, Thomas. 2004. Speaking of sound language and the professionalization of sound-recording engineers. *Social Studies of Science* 34 (5): 733–758.

Pritchard, Sara B. 2011. *Confluence: The Nature of Technology and the Remaking of the Rhône*. Cambridge, MA: Harvard University Press.

Pritchard, Sara B. 2012. From hydroimperialism to hydrocapitalism: "French" hydraulics in France, North Africa, and beyond. *Social Studies of Science* 42 (4): 591–615.

Putnam, Robert D. 1993. *Making Democracy Work: Civic Traditions in Modern Italy*. Princeton, NJ: Princeton University Press.

Quine, Willard Van Orman. 1961. *From a Logical Point of View*. Cambridge, MA: Harvard University Press.

Ravetz, Jerome R. 1971. *Scientific Knowledge and Its Social Problems*. Oxford: Clarendon Press.

Renzi, Barbara Gabriella, and Giulio Napolitano. 2011. *Evolutionary Analogies: Is the Process of Scientific Change Analogous to the Organic Change?* Newcastle upon Tyne: Cambridge Scholars Publishing.

Rosen, Paul. 1993. The social construction of mountain bikes: Technology and postmodernity in the cycle industry. *Social Studies of Science* 23 (3): 479–513.

Ross, Andrew. 1991. *Strange Weather: Culture, Science, and Technology in the Age of Limits*. London: Verso.

Rottenburg, R. 2009. *Far-Fetched Facts: A Parable of Development Aid*. Cambridge, MA: MIT Press. Originally published in German as *Weit hergeholte Fackten. Eine Parabel der Entwicklungshilfe* (Lucius & Lucius, 2002).

Russell, Stewart. 1986. The social construction of artefacts: A response to Pinch and Bijker. *Social Studies of Science* 16 (2): 331–346.

Sabel, Charles F. 2002. Diversity, not specialization: The ties that bind the (new) industrial district. In *Complexity and Industrial Clusters*, ed. Alberto Quadrio Curzio and Marco Fortis, 107–122. New York: Physica.

Sabel, Charles, and Jonathan Zeitlin. 1985. Historical alternatives to mass production: Politics, markets, and technology in nineteenth-century industrialization. *Past and Present* 108:133–176.

Sacks, Harvey. 1972. An initial investigation of the usability of conversational data for doing sociology. In *Studies in Social Interaction*, ed. David Sudnow, 31–74. New York: Free Press.

Sacks, Harvey. 1995. *Lectures on Conversation*. Oxford: Blackwell.

Sacks, Harvey, Emanuel Schegloff, and Gail Jefferson. 1974. A simplest systematics for the organization of turn-taking for conversation. *Language* 50 (4): 696–735.

Sayes, Edwin. 2014. Actor-network theory and methodology: Just what does it mean to say that nonhumans have agency? *Social Studies of Science* 44 (1): 134–149.

Scott, Janny. 1996. Postmodern gravity deconstructed, slyly. *New York Times*, May 18.

Segerstråle, Ullica Christina Olofsdotter, ed. 2000. *Beyond the Science Wars: The Missing Discourse about Science and Society*. Albany: SUNY Press.

Shapin, Steven. 1975. Phrenological knowledge and the social structure of early nineteenth-century Edinburgh. *Annals of Science* 32 (3): 219–243.

Shapin, Steven. 1979. The politics of observation: Cerebral anatomy and social interests in the Edinburgh phrenology disputes. In *On the Margins of Science: The Social Construction of Rejected Knowledge*, ed. Roy Wallis, 139–178. Keele: University of Keele.

Shapin, Steven. 1981. Of gods and kings: Natural philosophy and politics in the Leibniz–Clarke disputes. *Isis* 72:187–215.

Shapin, Steven. 1984. Talking history: Reflections on discourse analysis. *Isis* 75 (1): 125–130.

Shapin, Steven, and Simon Schaffer. 1985. *Leviathan and the Air-Pump: Hobbes, Boyle, and the Experimental Life*. Princeton, NJ: Princeton University Press.

Silverstone, Roger. 1994. *Television and Everyday Life*. London: Routledge.

Silverstone, Roger, and Eric Hirsch, eds. 1992. *Consuming Technologies: Media, and Information in Domestic Spaces*. London: Routledge.

Sismondo, Sergio. 2010. *An Introduction to Science and Technology Studies*, 2nd ed. Oxford: Blackwell. (1st ed. 2004.)

Sismondo, Sergio. 2015. Ontological turns, turnoffs, and roundabouts. *Social Studies of Science* 45 (3): 441–448.

Smith, Mark. 2012. The garden in the machine: Listening to early American industrialization. In *The Oxford Handbook of Sound Studies*, ed. Trevor J. Pinch and Karin Bijsterveld, 39–57. New York: Oxford University Press.

Sokal, Alan D. 1996. Transgressing the boundaries: Toward a transformative hermeneutics of quantum gravity. *Social Text* 46–47:217–252.

Sokal, Alan D. 2010. *Beyond the Hoax: Science, Philosophy, and Culture*. Oxford: Oxford University Press.

Sokal, Alan D., and Jean Bricmont. 1997. *Impostures Intellectuelles*. Paris: O. Jacob.

Sokal, Alan D., and Jean Bricmont. 1999. *Fashionable Nonsense: Postmodern Intellectuals' Abuse of Science*. New York: St. Martin's Press.

Sterne, Jonathan. 2001. Mediate auscultation, the stethoscope, and the "autopsy of the living": Medicine's acoustic culture. *Journal of Medical Humanities* 22 (2): 115–136.

Sterne, Jonathan. 2003. *The Audible Past: Cultural Origins of Sound Reproduction.* Durham, NC: Duke University Press.

Sterne, Jonathan. 2012a. *MP3: The Meaning of a Format.* Durham, NC: Duke University Press.

Sterne, Jonathan, ed. 2012b. *The Sound Studies Reader.* New York: Routledge.

Suchman, Lucy. 2007. Anthropology as "brand": Reflections on corporate anthropology. Paper presented at the Colloquium on Interdisciplinarity and Society, Oxford University, Feb. 24, 2007.

Sudnow, David, ed. 1972. *Studies in Social Interaction.* New York: Free Press.

Thompson, E. A. 2002. *The Soundscape of Modernity: Architectural Acoustics and the Culture of Listening in America, 1900–1933.* Cambridge, MA: MIT Press.

Thrift, Nigel. 2007. *Non-Representational Theory: Space, Politics, Affect.* London: Routledge.

Tosoni, Simone. 2003. Il Sapere Relazionale: Scenari di Interazione nella Contrattazione. In *La Lingua del Tumulto. Un'Archeologia dei Saperi di Borsa, edited by Ruggero Eugeni and Nevina Satta,* 131–183. Milano: Scheiweller.

Tosoni, Simone. 2007. Insicurezza e Stigma. Vivere nello Spazio Altro dei Quartieri Sensibili. In *La Città Abbandonata,* ed. Mauro Magatti, 345–410. Bologna: Il Mulino.

Tosoni, Simone. 2015. Addressing "captive audience positions" in urban space: From a phenomenological to a relational conceptualization of space in urban media studies. *Sociologica* 3. http://www.sociologica.mulino.it/journal/article/index/Article/Journal:ARTICLE:901 (last accessed Apr. 12, 2016).

Tosoni, Simone, and Matteo Tarantino. 2013. Space, translations, and media. *First Monday* 18 (11). http://firstmonday.org/ojs/index.php/fm/article/view/4956 (last accessed Apr. 12, 2016).

Travis, George D. L. 1981. Replicating replication? Aspects of the social construction of learning in planarian worms. *Social Studies of Science* 11 (1): 11–32.

Travis, George D. L. 1987. Memories and molecules: A sociological history of the memory transfer phenomenon. PhD thesis, University of Bath.

Traweek, S. 1988. *Beamtimes and Lifetimes: The World of High Energy Physicists.* Cambridge, MA: Harvard University Press.

Turkle, Sherry, ed. 2007. *Evocative Objects: Things We Think With.* Cambridge, MA: MIT Press.

Vannini, Phillip, ed. 2009. *Material Culture and Technology in Everyday Life: Ethnographic Approaches*. New York: Peter Lang.

Vaughan, Diane. 1996. *The Challenger Launch Decision: Risky Technology, Culture, and Deviance at NASA*. Chicago: University of Chicago Press.

Verbeek, Peter-Paul. 2005. *What Things Do: Philosophical Reflections on Technology, Agency, and Design*. University Park: Pennsylvania State University Press.

Verbeek, Peter-Paul. 2006. Materializing morality: Design ethics and technological mediation. *Science, Technology, and Human Values* 31 (3): 361–380.

Verbeek, Peter-Paul. 2011. *Moralizing Technology: Understanding and Designing the Morality of Things*. Chicago: University of Chicago Press.

Von Hippel, Eric. 1976. The dominant role of users in the scientific instrument innovation process. *Research Policy Research Policy* 5 (3): 212–239.

Von Hippel, Eric. 1988. *The Sources of Innovation*. New York: Oxford University Press.

Von Hippel, Eric. 2005. *Democratizing Innovation*. Cambridge, MA: MIT Press.

Wajcman, Judy, and Paul K. Jones. 2012. Border communication: Media sociology and STS. *Media, Culture, and Society* 34 (6): 673–690.

Wallis, Roy, ed. 1979. *On the Margins of Science: The Social Construction of Rejected Knowledge*. Keele: University of Keele.

Webster, Charles, ed. 1981. *Biology, Medicine, and Society, 1840–1940*. Cambridge: Cambridge University Press.

Weingart, Peter. 1974. On a sociological theory of scientific change. In *Social Processes of Scientific Development*, ed. Richard Whitley, 45–68. London: Routledge & Kegan Paul.

Whitley, Richard. 1972. Black boxism and the sociology of science: A discussion of the major developments in the field. *Sociological Review: Monograph* 18:61–92.

Whitley, Richard. 1974. *Social Processes of Scientific Development*. London: Routledge & Kegan Paul.

Whittington, William. 2012. The sonic playpen: Sound design and technology in Pixar's animated shorts. In *The Oxford Handbook of Sound Studies*, ed. Trevor J. Pinch and Karin Bijsterveld, 367–386. New York: Oxford University Press.

Winch, Peter. 1958. *The Idea of a Social Science and Its Relation to Philosophy*. London: Routledge.

Winner, Langdon. 1993. Upon opening the black box and finding it empty: Social constructivism and the philosophy of technology. *Science, Technology, and Human Values* 18:362–378.

Wise, Norton. 1996. The enemy without and the enemy within: A review of Gross and Levitt's "Higher Superstition." *Isis* 87 (2): 323–327.

Wittgenstein, Ludwig. 1956. *Bemerkungen Über Die Gründlagen Der Mathematik— Remarks on the Foundations of Mathematics*. Oxford: Blackwell.

Woolgar, Steve. 1982. Laboratory studies: A comment on the state of the art. *Social Studies of Science* 12 (4): 481–498.

Woolgar, Steve. 1988a. *Knowledge and Reflexivity: New Frontiers in the Sociology of Knowledge*. London: Sage.

Woolgar, Steve. 1988b. Reflexivity is the ethnographer of the text. In *Knowledge and Reflexivity: New Frontiers in the Sociology of Knowledge*, ed. Steve Woolgar, 14–36. London: Sage.

Woolgar, Steve. 1991. The turn to technology in social studies of science. *Science, Technology, and Human Values* 16:20–50.

Woolgar, Steve. 1993. What's at stake in the sociology of technology? A reply to Pinch and to Winner. *Science, Technology, and Human Values* 18 (4): 523.

Woolgar, Steve, and Javier Lezaun. 2013. The wrong bin bag: A turn to ontology in science and technology studies? *Social Studies of Science* 43 (3): 321–340.

Woolgar, Steve, and Javier Lezaun. 2015. Missing the (question) mark? What is a turn to ontology? *Social Studies of Science* 45 (3): 462–467.

Zaloom, Caitlin. 2006. *Out of the Pits: Traders and Technology from Chicago to London*. Chicago: University of Chicago Press.

Index